The Localization Reader

The Localization Reader

Adapting to the Coming Downshift

edited by Raymond De Young and Thomas Princen

The MIT Press
Cambridge, Massachusetts
London, England

For information about special quantity discounts, please email special_sales@mitpress.mit.edu

This book was set in Sabon by Toppan Best-set Premedia Limited. Printed and bound in the United States of America.

Library of Congress Cataloging-in-Publication Data
The localization reader : adapting to the coming downshift / edited by Raymond De Young and Thomas Princen.
p. cm.
Includes bibliographical references and index.
ISBN 978-0-262-01683-4 (hardcover : alk. paper) — ISBN 978-0-262-51687-7 (pbk. : alk. paper) 1. Sustainable living. 2. Environmentalism. 3. Energy conservation. 4. Energy consumption. I. De Young, Raymond, 1952– II. Princen, Thomas, 1951–
GE196.L63 2012
333.72—dc23

2011024378

10 9 8 7 6 5 4 3 2 1

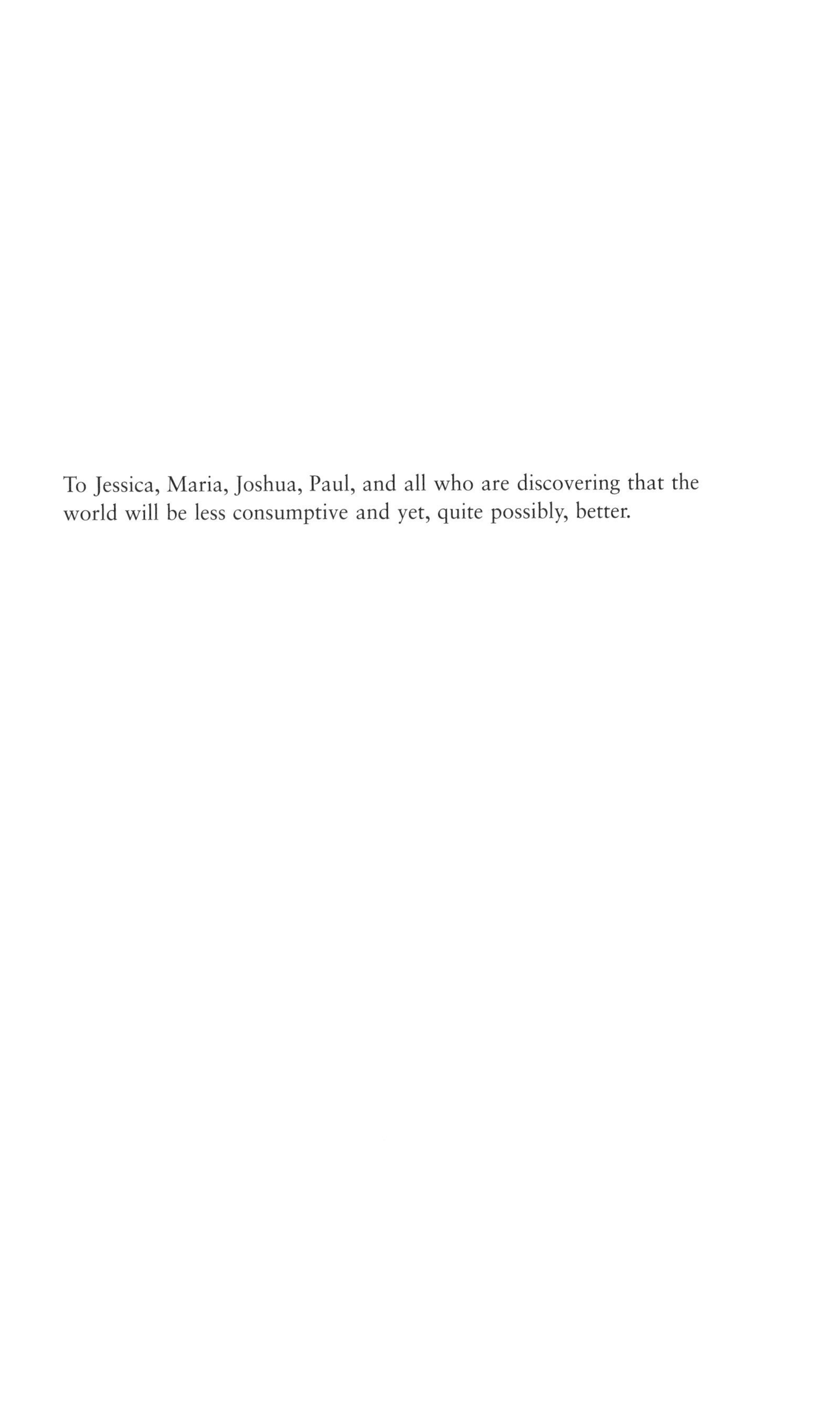

To Jessica, Maria, Joshua, Paul, and all who are discovering that the world will be less consumptive and yet, quite possibly, better.

Contents

Philosophies of localization affirm the possibility of flourishing while living with less and staying within natural limits. The arguments are clear and have been around a long time, but have been overshadowed by a consumerist culture, a focus on growth, and a cultural presumption that bigger and faster is always better.

act effectively. Some tools for making fundamental change already exist; others will certainly be needed.

Preface

Humans have always faced biophysical constraints. They could be ignored, though, for the past century or so, a brief, one-time era of fossil fuel abundance. This is no longer possible. Now the questions before us are these: How should societies respond to reemerging and unavoidable biophysical constraints? How can they transition in ways that are peaceful, democratic, just, and environmentally resilient? How might they craft a society that lives well and well within the limits of this single planet?

This book seeks to facilitate a conversation about these and other questions. Its focus is the response to a new era in the relationship between humans and the finite planet they inhabit. We believe that individuals and communities will enter this conversation, sometimes eagerly, sometimes reluctantly, as their outlook shifts from an expectation of ever-increasing material and energy abundance to one of limited resource availability. The conversation is needed, even overdue, but it will be hard. This collection of readings seeks to make this difficult conversation a bit less so.

We predict that the adaptation of human behavior and social organization to a changed worldview will proceed along two parallel, sometimes crossing paths. One path is on-the-ground practices, policies, and behaviors that exhibit an affirmative reframing of humans' relationships to one another and the environment. Existing examples of this include transition town projects in the United Kingdom and elsewhere, local and slow food movements, as well as initiatives in permaculture, ecovillages, and simple living. These efforts are experimental. Some will succeed, some will fail, but the practical knowledge gained will accumulate. Such knowledge is already disseminating widely, from one locality to another, across countries and around the globe.

The second path builds in part on these many small experiments and their accumulating knowledge. But it aims to step back and understand the ongoing transition within a larger context, a context that includes the transition's historical, political, behavioral, and biophysical antecedents and its predictions about a future with less material and energy. Constructing a framework for this context is the purpose of this book.

The practitioners—the on-the-ground experimenters and implementers—and the framers share at least one concern: as people come to understand the depth and breadth of the changes—some inevitable, some chosen—they are likely to find the transition unnerving. Facing a new reality often is. So it is important for all social change agents, while being upfront about the challenges, to also resist the temptation to use fear, to conjure up gloom-and-doom scenarios, or to assume that people must be coerced or incentivized before they will change. Using negative approaches, an all-too-common strategy in the environmental arena, often causes people to dismiss both the message and the messenger. Rather, we believe that it is incumbent on everyone to avoid overlooking (however unintentionally) the many positive changes that are already happening and to keep in mind that many more are likely to emerge.

In this book, we identify two ways social change agents can avoid overlooking or dismissing the positive. The first is to provide a framework that shows how a rapid transition can be both necessary and desirable and how it is that pragmatic plans for a transition already exist. The second is to document existing practices of localization. Early on the book outlines a scenario where biophysical constraints are taken as given. Then it describes current opportunities and positive efforts to respond thoughtfully to this plausible scenario. The book is an outline for crafting a positive response based on principles rather than a how-to guide for initiating a localization movement.

At the same time that we point to the positive, we must raise a caution. Localizers cannot assume that, based only on ideas and arguments like those presented in this book, people will readily abandon the many perceived benefits of industrial society. Localizers can assume, however, that as the unsustainable nature of high-consumption societies becomes obvious, people from all walks of life will be looking for practical models of how to live on less. We believe that building those models conceptually and on the ground is the localizers' main task.

Finally, we acknowledge that no one can be certain what principles, policies, and behaviors will work best in the coming decades. The conversation we wish to facilitate is necessarily a futuring exercise, a thought experiment that explores sensible responses to a resource-constrained future. Localization is only one such response but, as this book suggests, it is a highly plausible one. That said, we believe no single approach will fix things, everywhere, forever, or for all people. Thus, just as many experiments are already occurring, some hidden in plain sight, many more must be started. Indeed, a culture of small experiments must be fostered.

This book provides a context for such a culture. It aspires to give credibility to the practical experiments, join the practitioners in a common effort to point ahead, and begin a transformation as momentous as any humanity has ever faced. A transition will happen, but, we believe, it must start now to be positive. What is more, it is beginning, one store, one farm, one factory, one project, and one community at a time. This book offers ways of making it a positive change, a change that responds to inevitable descent in energy and material availability through a process we call localization.

Acknowledgments

This book grew out of conversations in a quiet courtyard, which led to a series of presentations and seminars. The seminar members—from disciplines such as environmental psychology, institutional analysis, resource policy, business, education, architecture, and planning—played a central role in framing this work, not to mention in finding and sorting through the material that follows. Without the questions, suggestions, and insights of the members, this collection could not have happened.

We extend special gratitude to James Crowfoot, who participated in the early stages of this project and guided us throughout. We also thank the contributors, both for permission to include their work and for the enthusiasm so many expressed for the concept of localization. Three anonymous reviewers provided useful feedback. Chris Stratman helped tame the early chaos of collecting the many pieces. We thank Nicole Premo, whose professionalism and attention to detail made this a better book. Clay Morgan applied a keen editor's eye in the overall shaping of the volume, and Sandra Minkkinen expertly shepherded it through production.

Finally, in an attempt to practice what the book promotes, we have committed all royalties to two community organizations that, from our personal experience, exemplify localization. Growing Hope is an organization dedicated to helping people improve their lives and communities through gardening, healthy food access, and local food security (www.growinghope.net) and People's Food Co-op has long sought to feed a community with wholesome food and good work (www.peoplesfood.coop).

Raymond De Young
Thomas Princen

Introduction

Localization is a process of social change pointing toward localities. Its primary concern is how to adapt institutions and behaviors to live within the limits of natural systems. In a localizing process, people's attention is focused on everyday behavior within place-based communities. At the same time, because localities are interdependent across scales, localization has regional, national, and international dimensions. Ultimately, localization's high-level goals are increasing the long-term well-being of people while maintaining, even improving, the integrity of natural systems, especially those that directly provide physical sustenance.

Localization is not strictly about the local and it is not to be confused with a narrowly focused localism. Nor is localization simply globalization in reverse. Rather, as overextended economies and resource extraction efforts exhaust themselves, we foresee industrialized societies experiencing a shift from the centrifugal forces of globalization (concentrated economic and political power, cheap and abundant raw materials and energy, intensive commercialization, displaced wastes, and abstract forms of communication) to the centripetal forces of localization (widely distributed authority and leadership, more sustainable use of natural energy sources and materials, personal proficiency, and community self-reliance).

Localization is a logical outgrowth of the end of a historically brief period, one that saw plentiful raw materials, highly concentrated and inexpensive energy sources, and an abundance of liquid fossil fuels whose wastes could disperse into the atmosphere, oceans, and soils without monetary costs. That period is coming to an end. How societies and individuals respond to this fact is one of the defining questions of our time. We presume in this book that people in all walks of life, in all positions of influence, will be asking not *if* localization will occur but how it *should* proceed. People will intuitively see that localization can be a force for good (e.g., healthy food, less anxiety, more neighborliness) or a force for evil (e.g., anarchy, warlords, survivalists, food deserts). In this book, we call the former *positive localization* and the latter *negative localization*.

The world is facing multiple challenges, each capable of shaking the very foundations of modern civilization. The central role that localization will soon play can be understood by briefly outlining the current global situation.

Climate Disruption

Climate disruption, once a mere hypothesis, is now empirically established. Through the efforts of the Intergovernmental Panel on Climate Change (IPCC) and other bodies, the science is clear: profound changes in the earth's thermal patterns are occurring and more will occur this century. Recent updates suggest that what were recently deemed worst-case and distant scenarios are now happening. Some believe that a global carbon management regime will eventually emerge out of the Rio/Kyoto/Copenhagen/Cancun negotiating process, but few are optimistic enough to suggest that such efforts will allow modern society to return to a preindustrial climate state. In fact, a realistic dose of pessimism has some groups promoting efforts to cope with, rather than just mitigate, climatic change. In either case, be it ruthless mitigation or revolutionary adaptation, high-consuming societies will have to operate on dramatically less material and energy in the foreseeable future. For that, we surmise, they will localize, ready or not.

Peaking of Fossil Fuels

A second challenge involves global energy dynamics. Fossil fuels (i.e., solar energy stored eons ago as hydrocarbon compounds) are the lifeblood of industrial civilization. The fact that this carbon store is finite is unquestioned, as is the empirical fact that eventually the rate of production from a reservoir *peaks*. For each reservoir, a maximum rate of extraction is reached after which production may plateau for a while but inevitably declines. Sometimes the decline is drawn out, sometimes abrupt. The same thing happens in the aggregate—that is, the production of multiple reservoirs also peaks. The implications of fossil fuel production peaking on a global scale are vigorously debated, albeit by a relatively small number of experts. The high emotions, competing competencies, and huge stakes that play out in this debate (as well as the astonishing political and media silence) make it hard to assess the urgency of fossil fuel peaking. It is even more difficult, but no less important, to find discussions about appropriate responses.

Nevertheless, agreement on one thing is emerging: soon the global production rate of liquid fossil fuels will peak, with other fuels and materials following suit soon after. Exactly when these peaks occur is much less important than the fact that they will occur. In fact, given how abruptly some reservoirs will drop in production, debating the exact timing can be a dangerous distraction. The task now is to prepare our response. It is to make plans for living—and thriving—while high-consuming societies descend the far side of "Hubbert's peak" (see Hubbert, chapter 1, this volume).

The Role of Technology

The energy crisis has sparked a profusion of technological improvements and publicly funded initiatives, with the great bulk aimed at maximizing efficiency. A century of technological modernization, including innovations that have helped propel dramatic and absolute increases in production and consumption, lead many people to be optimistic, even complacent, about finding technological solutions to all current and future dilemmas. We are less optimistic and we counsel against complacency.

First, technology cannot create new energy sources; it only transforms existing sources into forms more useful to modern society.

Second, some energy-saving technologies can paradoxically contribute to an increase in resource use—for example, fuel-efficient engines are made bigger and put into heavier vehicles, which are driven further and faster; electricity-efficient light bulbs are doubled and tripled to provide increased illumination.[1] Despite all the efficiencies gained, innovations in technology, policy, and practice have not reduced society's overall consumption of resources.

Third, even if society is temporarily reinvigorated by a newly transformed energy source (e.g., biofuels), as it grows, that source will eventually exhibit diminishing returns. In fact, stepping back for a moment and viewing this entire process, one can conclude that sociotechnical problem solving itself may be subject to diminishing returns (see Tainter, chapter 3, this volume). Supporting this conclusion is research suggesting that, to keep urban civilization going—that is, growing materially—innovations or adaptations must emerge at an accelerating rate to avoid stagnation or collapse. Yet, an ever-faster rate of discovery and implementation reaches a natural and completely unavoidable limit.[2]

New technologies might materialize in time and at a scale to ease the transition to noncarbon energy sources. But they only buy time for the transformation to a postcarbon world, a world with less net energy (see Hubbert, chapter 1, and Dadeby, chapter 2, this volume). Therefore, it is sensible to be somewhat pessimistic about the likelihood of technological innovations occurring just-in-time to ensure a smooth transition. And it is prudent to assume that, because no technology can overturn the biophysical laws of nature, other approaches are needed. It behooves us all to prepare for life with a lot less.

Premise

These observations—that climate disruption is occurring and will only intensify, that energy and material production will peak and then decline, and that technological innovation will help ease the transition but will not fundamentally change it—are what initially motivated us to teach the seminar from which this book emerged. They form the premise of this book. In designing the seminar we deliberately avoided dwelling on these facts—they are well discussed by others. Rather, we focused on the response. So, at this point it will be useful to state, plainly and briefly, the major components of the premise of this book:

1. Modern industrial society is facing a new biophysical reality, one that involves an inevitable decrease and, eventually, a leveling of material and energy availability at the same time that the consequences of past consumption must be addressed. This reality will negatively affect essential services and social institutions (e.g., food systems, health provision, mobility, banking).
2. These circumstances and ensuing effects are, like gravity, not negotiable. They are not altered by political debate or market forces, nor will denial or inattention make them disappear.
3. Conventional policy tools (e.g., pricing and markets, technological innovation) will not be up to the task.
4. The speed and suddenness of change mean the operative term is *response*. It requires preparation so as not to be taken by surprise. To prepare reasonable responses, this book sketches plausible future scenarios, each of which assumes that the coming downshift is inevitable. What is not inevitable, however, is the nature of those responses.

These, then, are the four key parts of the premise of the book. They can also be framed as a prospect:

1. Without a plan or a process, society risks a rapid, chaotic descent into a hyperlocal existence, what we characterize as negative localization.
2. Positive localization, in contrast, is a process for creating and implementing a response, a means of adapting institutions and behaviors to living within the limits of natural systems. Place-based localization includes institutions at the regional, national, and international levels.
3. Localization is not an outcome or end state to pursue. Rather it is a way of organizing and focusing a process of transition. It is, arguably, a process already underway, but one that should be accelerated while options still exist.

Transitioning While There Are Surpluses

The likelihood of a long descent in material and energy abundance, along with the disruption of ecosystems, including the climate, suggest that the transition to sustainable living should start soon, while surpluses of social, ecological, and economic capital remain. Unlike short-term emergencies where life eventually returns to normal, these conditions will lead to a new normal. Arguably, that new normal is already emerging, driven in part by biophysical necessity (e.g., glacier-dependent populations adapting to diminishing water supplies) and in part by creative, anticipatory response (see part II). Even an optimistic view of the promises of global management and technological innovation leads to the realization that, once energy and material limits are reached, we will still require place-based solutions; proactive preparation is thus desirable. Furthermore, the behavioral and institutional responses must be durable. Lapses back to spendthrift ways, borrowing heavily from the future, and consuming now while hoping that some discovery will return us to the old days of unlimited growth will only exacerbate the transition, enabling the negative forms of localization to establish themselves.

The above recitation of trends and responses is the sobering part of this introduction. Now for the hopeful part, indeed the exciting part.

Distinguishing Localization

From where we sit, it seems that the approaches for dealing with industrial society's fossil fuel dependence and the resulting environmental disruptions come from two different worlds—the world of experts and the world of communities. In the experts' world, centralized power, specialized knowledge, and top-down, elite-driven global management are taken as self-evident. In this world, people—that is, the masses, the unsophisticated—are the target of global management plans and the question is "How do we get people to behave properly?" The "we" are the elites who have the solutions and the "people" are the source of the problem whose behavior must be shaped. In this world, the experts have trouble seeing any utility in something like localization.

In the world of communities, the self-organizing capacities of ordinary people are paramount. Here, local knowledge and group efficacy lead to locally compatible solutions. Peoples' behavior must still change, but citizens define the problem and become the source and disseminator of its solutions. In this world, citizens working in communities see localization as a useful tool for crafting a durable and just society.

While the experts' world renders small-scale, low technologies trivial, the world of communities values technology at an appropriate scale. Where the experts' world evaluates small businesses and family farms as inconsequential, the world of communities understands the resilience of locally owned, independent enterprises. Where the world of experts dismisses grassroots organizing as ineffective, the world of communities understands the staying power of participatory democracy.

In practice, the process of positive localization has three salient features:

1. *It doesn't plead.* Localization doesn't beg or try to coerce people to act right. It doesn't say people must appreciate nature, consider future generations, or save the environment. All those behaviors may be desirable, but localization presumes that the consequences of growing populations, increasing consumption rates, and past and current emissions are such that high-consuming societies *will* be adapting whether they are environmentally enlightened or not. What is more, behavior change may precede any change in attitudes, values, or worldviews.

2. *It holds little faith in centralized approaches.* Top-down, command-and-control, get-the-incentives-right, correct-market-imperfections approaches, which might be lumped under the term *global management*, contain implicit models of social change that privilege centralized power. What is more, these models do not draw on the extensive research that shows how to engage people in reasonable behavior. Localization builds on such research. It presumes that people are more motivated and content when they are solving problems and that they are perfectly capable of organizing themselves to live within immutable biophysical limits (see part IV). Participatory democracy may be difficult—even tedious, frustrating, and

inefficient—but it is an essential component for creating meaningful lives and true communities, especially under the conditions posited here (see part V).

3. *It is affirming.* Localization affirms self-organization, self-reliance, self-limitation, and self-rule. It is not protest, not antiglobalization, not antitechnology, and not antimodern. It assumes everyone has the ability to contribute to a solution. While no one group is required to complete the work, neither are any groups free to avoid doing their share. It seeks appropriate technologies and well-regulated markets that build in responsibility among producers, consumers, investors, and regulators.

An Affirmative Direction with Embedded Benefits

As difficult as it is to change institutions and behavior and as inconvenient as the changes will be, an affirmative approach has embedded benefits. These positive consequences of localization easily go unnoticed, especially when, understandably, people are not careful and patient observers. But some people have begun to adapt in place. If one looks closely, vibrant examples of intrinsically satisfying efforts at simple living, shared transportation, food provisioning, local finance, and cooperative enterprises are springing up all over, sometimes in the unlikeliest of places. And they are doing so with little if any support from the dominant institutions of government and commerce.

Embedded benefits have two features. One is that they are easily obscured under the heavy covers of commercialism and consumerism. The benefits of localization exist but they are hard to see. In this book, we aspire to pull back the covers and show these benefits.

Embedded benefits are also found in the process of problem solving. The challenges of the coming transition are great but humans are nothing if not great problem-solvers. They are at their best when they help themselves and help others, when they are called on to be creative and self-directed, and when they are tackling problems that are challenging, genuine, and meaningful. Human ingenuity, long aimed at crafting an industrial society, must now be aimed at crafting a durable civilization. The creative effort contains its own rewards.

In the classroom we have found students to be great sources of such creativity. We are confident others will join, creating a *conversation* about and a *practice* of positive localization. They will reorient daily and public life back to person-to-person and human-to-nature partnerships, all at an appropriate pace and a human scale (see chapter 11, this volume).

We say this but we must be cautious in our use of language. "Back to" in this book does not mean reversion to a primitive state; it does not mean going backward. It does mean finding cultural assets that worked in less consumptive times and adapting them to current times; it means keeping the good of the present and eliminating the bad. What's more, it does not mean starting a new environmental movement. Unlike traditional activism responding to environmental insults, energy shortages, and climate disruption, we do not aim to spark protests against government and industry, denounce the excesses of consumerism, or debate the pros and cons of carbon trading and carbon taxes. Localization is not a revolution in the streets, or a new strategy for corporate and NGO headquarters. Rather, it is an affirmative social trend, driven by biophysical realities and accepting of the innate human inclinations for self-provisioning and commitment to place.

The Readings

While energy is one key resource driving the current debate, this book does not emphasize energy policy and planning; many excellent books exist on that topic. Nor does it dwell on the gloom-and-doom scenarios emerging from the debates over peak oil and climate disruption. We focus instead on understanding the emerging transition while providing guidance toward a wholesome, just, and resilient version of that transition.

The collected works guide readers through the nuances of the topic—for example, localization is not simply "the local" or the reverse of globalization; the transition will be demanding but not necessarily destructive, possibly uplifting; historical precedents do exist although a global resource reduction at the level being contemplated is unprecedented, making discovery needed. People need to be engaged in a process, the details of which cannot be worked out by others, certainly not by decision makers far removed from people's everyday existence.

So we begin these readings with the *context* of localization. The early chapters lead to one unavoidable conclusion: society will be making a fundamental transition away from fossil fuels, willing or not, ready or not. While such a future may be frightening to many, human history suggests that we have adequate familiarity with such changes. In fact, some of these readings show that humans have longstanding decentralist tendencies that have served them well. Accepting the inevitability of transition, however, is separate from knowing the transition's trajectory. The collected material will help readers envision a variety of possible paths. Some will be quite familiar, although none are extrapolations from present high-consuming life patterns.

The middle chapters of the book outline ways to self-organize, self-govern, and self-provision while material and energy abundance is diminishing. Some of the readings look to the agrarian past for advice on how to live in partnership with natural systems, while others suggest new ways to manage the exchange of goods and money and the structures of ownership. This section also includes readings that suggest the deep fulfillment that such an existence can provide.

Human societies were once organized locally but this is no longer the dominant pattern. A new pattern is possible, indeed imperative, but it requires effective adaptation and strategic management. The later chapters of the book take up this issue by exploring human needs and strengths and the conditions under which humans effectively problem solve. The readings then outline approaches for working with innate human tendencies to initiate a societal transition to sustainable living. A key notion in these readings is that successful approaches will be those that enable people to discover for themselves how to transition well, rather than rely on others to plan and manage the transition for them.

An insight common to many of the readings is that localization is well underway. However, existing practices lack a framework for understanding and coping with declining net energy and other biophysical constraints. This book begins building such a framework, preliminary as it is. The framework assumes that a fundamental departure from recent life patterns has begun and that much about the transition will be hard. However, adjusting to low levels of consumption can be satisfying in a way present generations have forgotten or never experienced.

A final note on the readings: some are quite old. For example, Royce wrote his essay in 1908. It may be easy to dismiss such musings as irrelevant to the contemporary ecological predicament, a predicament that was not even on the horizon in the early 1900s. But we hope the reader will see that this and other readings were harbingers of the current predicament of unending growth on a finite planet. These readings help frame the issues in a context of long-term social change. In that historical trajectory we witness many failures, such as the nineteenth-century populism discussed by Royce. Such efforts were overwhelmed by larger forces, in particular by the great wealth and power afforded by industrialization, commercialization, and consumerism. These forces were fueled by biophysical circumstances that no longer exist: abundant fossil fuels, cheap to mine with wastes virtually costless to dispose of. So if in the past—at a time awash in cheap energy and endless waste sinks—there were compelling reasons to start a conversation about durable living, those reasons are all the more compelling now.

Positive localization is a dynamic, ongoing, and long-term process that can bring out the best in people. It is about directing the transition in a peaceful, just, psychologically enriching, and ecologically resilient way. It asks how new patterns of living can be fostered through creative exploration as attention shifts from the global and abstract to the close-at-hand and tangible. In short, the position of this book is that there is inevitability in diminishing material and energy availability but not in the way humans will adapt to this new reality.

In sum, our task in this book is to find new language that shines light on the localizing world. We recognize that the most powerful searchlights won't alter the view of those active in globalization. Indeed, they are not our audience. Our audience includes people who see a looming cliff but also see a rare opportunity for meaningful change. Our audience sees no future in endless material growth on a finite planet. In fact, they see the utter illogic in it. Our audience also includes people who value direct relations with others and nature, who find restraint and moderation satisfying, even uplifting. Our audience includes people who are *doing* localization already, or are contemplating doing it. We hope this book provides them support. We hope it helps them frame their good works as part of a larger, meaningful struggle. We hope to contribute to the legitimacy of such efforts so that, as localizers multiply, their work becomes the norm.

Notes

1. For an explanation of how efficiencies can increase resource use, see "Whose Ratios? From Technic to Rhetoric" in Thomas Princen, *The Logic of Sufficiency* (Cambridge, MA: MIT Press, 2005).

2. Luis M. A. Bettencourt, Jose Lobo, Dirk Helbing, Christian Kuhnert, and Geoffrey B. West, "Growth, Innovation, Scaling, and the Pace of Life in Cities," *PNAS* 104, no. 17 (2007): 7301–7306.

I

Drivers of Localization

This book's readings start with a somber tone. Global resource availability is approaching a plateau and is likely to soon begin a long descent. Declining energy availability, particularly liquid fossil fuels, threatens industrial society. This is not recent news. For some time now, analysts such as M. King Hubbert have rung alarms, showing the impossibility of exponential growth in resource extraction (chapter 1). Ecological economist Herman Daly has called unending growth an impossibility theorem. The implications for societies addicted to growth, however, are another matter.

The concept of peak oil is also relatively straightforward. It is not about pumping the last barrel of oil but rather about the rate at which oil can be extracted. After a period of ever-increasing extraction, the rate of production eventually peaks, followed by a decline. Hubbert's name is given to the curve describing this process. It applies to single reservoirs and, in general, to the aggregation of all reservoirs. The front side of Hubbert's peak—the growth side—is familiar to all. It is the implicit physical basis of nearly all we know about modern industrial society. It is the back side that gives analysts pause. Because the oil that is first pumped in a given region comes from high-quality reservoirs, near the surface, often under significant pressure and in easy-to-reach places, the remaining oil is harder and more expensive to extract and process. These latter sources come from deeper reservoirs, in inhospitable regions (e.g., Alaska's North Slope), far offshore, deep under the seabed (e.g., the Gulf of Mexico, the coast of Brazil), and from unconventional sources (e.g., tar sands, oil shale). These sources require advanced technologies, more capital investment, and, most significantly, *increased amounts of energy*, which, from a thermodynamic perspective, is the ultimate determinant of net energy availability. It takes energy to get energy and transform it into usable forms.

This pattern—pursue the easiest to get first—while perfectly sensible economically and technologically, means that, over time, the *quality of the resource extracted declines*—and it does so as inevitably as the seasons turn. It is a physical fact. No amount of capital, no level of technological sophistication, and no amount of political debate can change this fact. It means that less and less net fossil fuel energy remains for operating industrial civilization. Energy analyst Adam Dadeby explains how this works in terms of energy returned on energy invested (EROEI, chapter 2).

Anthropologist and historian Joseph A. Tainter applies a similar notion—diminishing marginal returns—to social organization (chapter

3). Societies tend to solve problems by increasing social complexity. They introduce more complicated technologies, create new institutions, add regulations, pile on bureaucratic layers, and use more specialists to gather and analyze yet more information. As societies do all this, the cost of each additional level of complexity comes to outweigh the benefits derived from that added complexity. These diminishing returns on investments eventually result in ineffective, counterproductive problem solving and, as a result, increased vulnerability to collapse. Collapse might be sudden but need not be total. It can simply be a return to an earlier, lower level of social complexity, a situation that social critic Ivan Illich implicitly argues for (chapter 4). Illich connects high levels of energy use to social inequity and a low quality of life. Thus, despite the gloom associated with the very idea of collapse, there is a hopeful side—with lower levels of energy use may come greater social equity—and, we surmise, more time spent in the community which can lead to increased social interaction, meaningfulness, and quality of life. For present purposes, namely the construction of a positive localization, it may be useful to avoid the term *collapse*, with its inherent negative connotation, and adopt a term like *downshift* (chapter 25).

Peak oil and collapse (or downshift) have their skeptics, much like climate change has its. Because these ideas have barely entered the public debate, those who object tend to demand irrefutable evidence for declining net energy or for the inevitability of diminishing returns on social complexity. Our position is that those who find peak oil and downshift plausible can devote their attention to one of two pursuits: countering the skeptics or helping to prepare for the downshift. This book chooses preparing. It is about framing a social trend that has only barely begun and needs help developing.

For this reason, we encourage the reader to proceed as if those who speak seriously about peak oil and collapse are right. If the "peakists" turn out to be wrong, the changes society adopts will likely be for the better anyway—living on less, in stronger communities, with more time, enhanced well-being, and better food. For guidance in preparing for the downshift, consider the following principles:

1. *Precautionary principle.* It is appropriate to take action on low-probability, high-impact events even in the absence of scientific certainty.

2. *No-regrets policymaking.* Policies with net benefits are sensible even if the initial assumptions prove to be wrong.

The skeptics will, at times, still need refuting. We suggest beginning by carefully examining their arguments and operating assumptions. Often these have an unstated premise that lacks plausibility—for instance, the possibility of endless resource extraction and consumption on a finite planet or never-ending substitutions afforded by new technologies. Then demand of the skeptics explanations that are grounded in the biophysical. Ask for evidence that people's sense of psychological well-being *requires* increasing per capita consumption. The lack of such evidence may not mute their skepticism, but it may free up localizers' energy to get on with the task of preparing for the downshift, our task in this book.

Contemplating dramatic change in civilization as we know it, whether a sudden collapse or a planned downshift, is admittedly uncomfortable. It should be done only if it serves a greater purpose. We pose these issues not to be pessimistic, not to scare people into action, not to break their denial and lethargy. Rather, we pose them because, in the first instance, the descent scenario is eminently plausible, albeit upsetting. In the second, we pose them because we are convinced fruitful change is possible: positive localization can happen.

Here is one final piece of advice about such matters from a colleague, psychologist Rachel Kaplan. She recommends practicing *anticipated regret*. Imagine a time after fossil fuel production has peaked and some of the worst-case scenarios of global climate disruption have occurred. What efforts might we then regret not having made today? What direction would we then regret not steering society in today? These initial readings begin this process of anticipated regret by outlining the drivers of localization, helping us imagine a future state, one we will need to respond to and shape.

1

Fossil Fuel Decline*

M. King Hubbert

Writing a half century ago, geoscientist M. King Hubbert begins the following essay with a simple but profound observation: for all but the last 150 years of history, humans have depended on renewable energy, and they will again soon. Cheap, dense, readily transportable, nonrenewable energy sources are expected to drastically decline (they may have started declining already) and will never be replaced; they are "absolutely exhaustible," assured by the very nature of geology. Although his data are out of date, and although scientists have learned much in the ensuing decades (about, for example, embodied energy and total reserves, not to mention climate disruption brought on by fossil fuel consumption), Hubbert's warning of "exponential growth as a transient phenomenon" remains: constant growth from finite resources is impossible. And his challenge is the same, only more urgent: we must figure out "how to make the transition from the precarious state that we are now in to this optimum future state by a least catastrophic progression." If we act "before unmanageable crises arise," Hubbert concludes, "there is promise that we could be on the threshold of achieving one of the greatest intellectual and cultural advances in human history." Notice that instead of using wording like decline, collapse, *or* go back, *Hubbert envisions a response to biophysical limits as an* advance.

In this bicentennial year of American history, it is useful for us to reflect that the two-hundred-year period from 1776 to 1976 marks the

*Hubbert, M. King. 1996. "Exponential Growth as a Transient Phenomenon in Human History." In *Valuing the Earth: Economics, Ecology, Ethics*, ed. Herman Daly and Kenneth Townsend, 113–126. Cambridge, MA: The MIT Press. Excerpts reprinted with permission from "Societal Issues, Scientific Viewpoints," by Margaret A. Strom. Copyright 1987, American Institute of Physics.

emergence of an entirely new phase in human history. This is the period during which our industrial civilization has arisen and developed. It is also the period during which there has occurred a transition from a social state whose material and energy requirements were satisfied mainly from renewable resources to our present state that is overwhelmingly dependent upon nonrenewable resources. In 1776 our material requirements for food, housing, clothing, and industrial equipment were principally satisfied by renewable vegetable and animal products. Nonrenewable mineral products, clay products, lime, sand, and metals, were used in such small amounts that the available supplies, at that rate of consumption, seemed almost inexhaustible.

The energy requirements two centuries ago were likewise met principally by renewable resources. Vegetable and animal products were used for food and warmth; human and animal labor and wind and water power for mechanical work. The only nonrenewable energy source then used was coal, which was consumed in such small amounts per year that the total supply at this rate would likewise have seemed almost inexhaustible.

During the ensuing two centuries, the development of the world's present highly industrialized society has occurred. . . .

The mining of coal as a continuous industrial enterprise began nine centuries ago near the town of Newcastle upon Tyne in northeast England. Annual statistics of world coal production are difficult to assemble before 1860, but by that time the annual production rate had reached 138 million metric tons per year. From the earlier history of coal mining and from scattered statistics it can be estimated that during the eight centuries from 1060 to 1860, the average growth rate in annual coal production was about 2.3 percent per year with an average doubling period of about 30 years. From this it can be estimated that the production of coal in 1776 was about 20 million metric tons. As of this year, the annual coal production rate has reached about 3.3 billion metric tons—a 165-fold increase during the two centuries since 1776. This has been at an average growth rate of about 2.55 percent per year, or an average period of doubling of 27 years. Although very small amounts of oil were produced in China and Burma at earlier times, the world's production of crude oil as a continuous industrial enterprise was begun in Romania in 1857 and in the United States two years later. . . . From 1880 until 1970 the growth in annual crude oil production increased smoothly and spectacularly. During this period the growth rate averaged 7.04 percent per year and the production rate doubled, on the average, every

9.8 years. The cumulative production also doubled about every 10 years. For example, the amount of oil produced during the decade 1960–1970 was almost exactly equal to all the oil produced from 1857 to 1960. . . .

In 1776 the world's human population was approximately 790 million. It has increased to 4.24 billion. The per capita consumption of iron in 1776 amounted to only 0.46 kilogram. This has increased to 132 kilograms, a 287-fold increase during these two centuries.

These . . . are . . . profound changes in human affairs . . . during the last two centuries. Our present concern, however, is with the future. Is it possible that such rates of growth can be maintained during the next two centuries, or do the industrial and demographic growth rates experienced during the last two centuries represent a transient and ephemeral epoch in the longer span of human history?

Answers to such questions can be obtained by considering the physical nature of various growth phenomena. Consider first a renewable phenomenon such as a food supply or the development of water power. Human food supply is derived almost entirely from plant or animal products. Therefore, the problem of the increase of food supply, or of the human population itself, reduces to the basic problem of how much the population of any biologic species can be increased on the earth. The basics of this problem are well understood. Biologists discovered a couple of centuries ago that the population of any biologic species, plant or animal, if given a favorable environment, will increase exponentially with time. That is, the population will double and redouble in the successive ratios of 1, 2, 4, 8, etc. during successive equal intervals of time. The period required for the population to double is different for different species. For elephants and humans the doubling period is a few decades, but for some bacteria it is as short as 20 minutes. Such a manner of growth obviously cannot continue indefinitely, but the significant question is this: About how many successive doublings on a finite earth are possible?

An approximate answer can be obtained when we consider the magnitudes obtained by successive doublings. Consider, for example, the classical problem of placing one grain of wheat on the first square of a chessboard, two on the second square, four on the third, and continuing the doubling for each successive square of the board. On the *n*th square—the last, or 64th—the number of grains will be 2^{n-1} or 2^{63}. How much wheat will this amount to? On the last square alone, the amount of wheat would be equal to approximately 1,000 times the world's present annual wheat crop; for the whole board, it would be twice this amount.

From one point of view this is merely a trivial problem in arithmetic; from another it is of profound significance, for it tells us that the earth itself will not tolerate the doubling of 1 grain of wheat 64 times.

Similar results are obtained when we consider the successive doublings of other biologic populations, or of industrial activities. The present world human population, were it to have descended from a single pair, say Adam and Eve, would have been generated by only 31 doublings, and a total of 46 doublings would yield a population density of one person per square meter over all the land areas of the earth. For an industrial example, the world's population of automobiles is also doubling repeatedly. If we apply the chessboard arithmetic to the automobile population, beginning with one car and doubling this 64 times, and then let the resulting quantity of automobiles be stacked uniformly on all the land surfaces of the earth, how deep would be the layer formed? About 1,200 miles or 2,000 kilometers.

From such considerations it becomes evident that the maximum number of doublings that any biological population or industrial component can experience is but a few tens. Therefore, any rapid rate of growth of such a component must be a transient phenomenon of temporary duration. The normal state of a biologic population, when averaged over a few years, must be one of an extremely slow rate of change—a near steady state. For example, the maximum possible number of times the human population could have doubled during the last million years is 31. Consequently, the minimum value of the average period of doubling during that time would have been 32,000 years, as contrasted with the present period of but 32 years.

The growth curve that characterizes a rapid increase of any biologic population or the exploitation of any renewable nonbiologic resource such as water power must accordingly be similar. . . . Beginning at a near steady state of a constant biologic population, or at zero as in the case of water power, the quantity considered increases exponentially—that is, it doubles in equal intervals of time—for a while, and then, in response to retarding influences, gradually levels off to a constant figure, characterized by a state of nongrowth.

In the biologic case the inverse can also happen. The disturbance may be unfavorable and the population may undergo a decline or a negative growth. It then must stabilize at a lower level, or else become extinct.

The long-term behavior of the exploitation of the nonrenewable resources such as the fossil fuels or the ores of metals differs from that of the renewable resources in a very fundamental respect. Nonrenewable resources are absolutely exhaustible. . . . The fuel is taken from the earth, and, in burning, is combined chemically with atmospheric oxygen. The materials, in the form of gaseous compounds CO_2, H_2O, and SO_2 and of asheous solids, remain on the earth, but the energy content released initially as heat, after undergoing successive degradations, eventually leaves the earth as spent low-temperature radiation. . . .

Thus the production history of the fossil fuels or of the metallic ores, in any given region and eventually for the entire earth, is characterized by a complete-cycle curve. . . . Production, beginning initially at zero, increases, usually exponentially, for a period. The growth rate then slows down and the rate of production passes one or more maxima and finally, as the resource becomes exhausted, goes into a long negative-exponential decline. . . . If from geological mapping, drilling, and other means, the total recoverable quantity of the resource considered can be estimated in the earlier stages of exploitation, the future of the production history can be estimated, because the complete-cycle curve can only encompass a total area corresponding to the estimate made.

By this means we are able to gain a reasonably reliable estimate of the time during which the fossil fuels can serve as major sources of the world's industrial energy. . . . Two aspects . . . are particularly significant, the approximate date of the peak rate of production, and the period of time during which the cumulative production increases from 10 to 90 percent of the ultimate cumulative production—the time span required to produce the middle 80 percent. For the high estimate of 7.6 trillion tons the estimated date of peak production would be about 150 to 200 years hence, and the time period required to produce the middle 80 percent would be that from about the year 2000 to 2300 or 2400. For the lower figure of 2 trillion tons of ultimate production the peak in the production rate would occur earlier, about the year 2100, and the middle 80 percent of the ultimate coal would be consumed during the approximately 200 years from about 2000 to 2200.

The estimate of the complete cycle for the world production of crude oil is . . . based upon an estimate of 2,000 billion barrels for the ultimate production, which is somewhat higher than the average of 15 estimates published since 1958 by international oil companies and leading international petroleum geologists. Using this figure, and assuming an orderly evolution of petroleum consumption, the peak in the production rate will probably occur during the decade 1990 to 2000, and the middle 80 percent will be consumed within the 60-year period from 1965 to 2025.

Hence, children born within the last 10 years will see the world consume most of its oil during their lifetimes. . . .

The known and estimated ores of [iron and] most other industrial metals, copper, tin, lead, zinc, and others, are in very much shorter supply, with the time until the peak production rate occurs and the time span for the middle 80 percent being measurable in decades rather than in centuries.

Three types of growth phenomena with which an industrial society must deal are shown graphically in figure 1.1. Here, the lower curve represents the rise, culmination, and decline in the production rate of any nonrenewable resource such as the fossil fuels or the ores of metals. The middle curve represents the rise and leveling off of the production of a renewable resource such as water power or a biological product. The third curve is simply the mathematical curve of exponential growth. No physical quantity can follow this curve for more than a brief period of time. However, a sum of money, being of a nonphysical nature and growing according to the rules of compound interest at a fixed interest rate, can follow that curve indefinitely.

In their initial phases, the curves for each of these types of growth are indistinguishable from one another, but as industrial growth approaches maturity, the separate curves begin to diverge from one another. In its present state the world industrial system has already entered the diver-

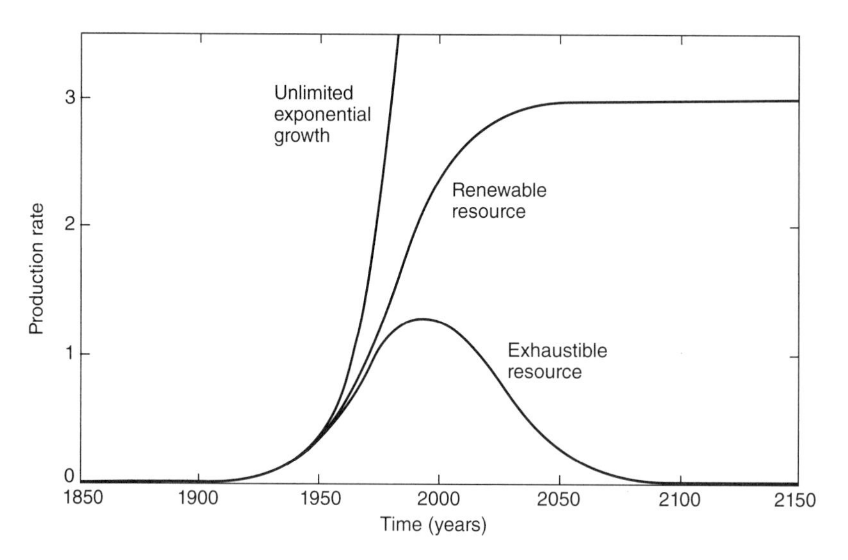

Figure 1.1
Three types of growth (Hubbert, 1974b, fig. 1)

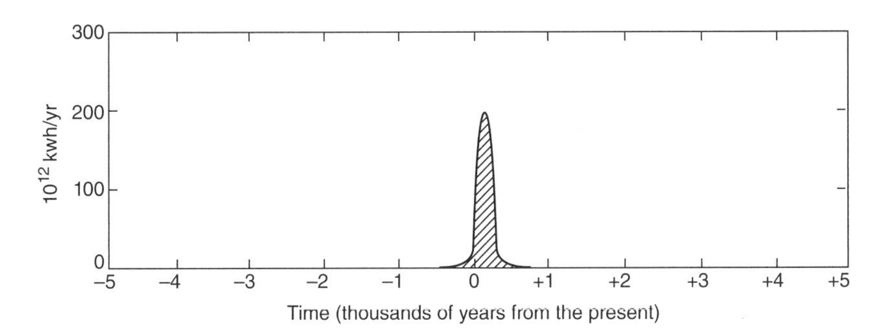

Figure 1.2
Epoch of fossil fuel exploitation in human history during the period from 5,000 years ago to 5,000 years in the future (Hubbert, 1974a, fig. 69)

gence phase of these curves but is still somewhat short of the culmination of the curve of nonrenewable resources.

A better appreciation of the brevity and exceptional character of the epoch of the fossil fuels can be gained if we view it in the perspective of a longer time span of human history that we have considered heretofore. In figure 1.2 the complete cycle of exploitation of the world's total supply of fossil fuels, coal and petroleum, is shown on a time scale extending from 5,000 years in the past to 5,000 years in the future. On such a scale, the Washington Monument–like spike in the middle of this range, with a middle 80 percent spread of about three centuries, represents the period of exploitation of the fossil fuels in the much longer span of human history. Brief as this period is, having arisen, as we have seen, principally within the last century, it has already exercised one of the most disturbing influences ever experienced by the human species in its entire biological existence.

The position in which human society now finds itself in this longer span of history is . . . a period of transition between a past characterized by a much smaller population, a low level of technology, of energy consumption, and of dependence upon nonrenewable resources, and by very slow rates of change, and a future also characterized by slow rates of change, but by means of utilization of the world's largest source of energy, that of inexhaustible sunshine, capable of sustaining a population of optimum size at a very comfortable standard of living for a prolonged period of time.

It appears therefore that one of the foremost problems confronting humanity today is how to make the transition from the precarious state that we are now in to this optimum future state by a least catastrophic progression. Our principal impediments at present are neither lack of energy or material resources nor of essential physical and biological knowledge. Our principal constraints are cultural. During the last two centuries we have known nothing but exponential growth and in parallel we have evolved what amounts to an exponential-growth culture, a culture so heavily dependent upon the continuance of exponential growth for its stability that it is incapable of reckoning with problems of nongrowth.

Since the problems confronting us are not intrinsically insoluble, it behooves us, while there is yet time, to begin a serious examination of the nature of our cultural constraints and of the cultural adjustments necessary to permit us to deal effectively with the problems rapidly arising. Provided this can be done before unmanageable crises arise, there is promise that we could be on the threshold of achieving one of the greatest intellectual and cultural advances in human history.

References

Hubbert, M. King. 1962. *Energy Resources: A Report to the Committee on Natural Resources.* National Academy of Sciences–National Research Council, Publication 1000-0.

Hubbert, M. King. 1972. "Man's Conquest of Energy: Its Ecological and Human Consequences." In *The Environmental and Ecological Forum 1970–1971,* ed. by Burt Kline, 1–50. U.S. Atomic Energy Commission Office of Information Services. Available as TID 25857, National Technical Information Service, U.S. Department of Commerce, Springfield, VA 22151.

Hubbert, M. King. 1974a. "U.S. Energy Resources: A Review as of 1972," Part 1. Committee on Interior and Insular Affairs, U.S. Senate, Serial no. 93-40 (92-75). U.S. Government Printing Office, 267 pp.

Hubbert, M. King. 1974b. "Statement on the Relations between Industrial Growth, the Monetary Interest Rate, and Price Inflation." Hearings before the Subcommittee on the Environment, Committee on Interior and Insular Affairs, House of Representatives, June 4, 6, and 10, July 19 and 26, 1974. Serial no. 93-55, pp. 58–77. U.S. Government Printing Office.

2

Energy Returned on Energy Invested*

Adam Dadeby

In this chapter, energy analyst Adam Dadeby reviews a decision-making tool rarely used in policymaking. Its simplicity should not distract from its profound implications, particularly as industrial civilization descends from the peak of resource availability that Hubbert describes. Energy returned on energy invested (EROEI) is a means of understanding why the depletion of nonrenewable resources poses such an enormous challenge to modern society. It demonstrates that we cannot grow our way out of the problem of fossil fuel dependence.

*Political scientist Thomas Homer-Dixon has argued that declining EROEI was one factor underlying the collapse of the Western Roman Empire in the fifth century CE (*The Upside of Down*, 2006, Knopf). To further understand the process behind the decline of civilizations, in the next reading, Joseph Tainter adds the complementary notion of diminishing marginal returns of societal complexity. Thus EROEI is useful for exploring our past, but EROEI is also a tool for exploring future scenarios, including durable living (see the box titled "Breaking the Fossil Fuel Habit").*

Energy returned on energy invested (EROEI or EROI) is a concept that mirrors the financial metric, return on investment (ROI). To make an energy gain or "profit," energy or work must be consumed or exerted.[1] The energy gain is often referred to as "net energy." EROEI is usually expressed as a ratio, or occasionally as a percentage. EROEI can also be represented diagrammatically in simplified form (figure 2.1). The energy referred to in EROEI can be that used to run technology, such as liquid

*This is the first publication of this article. Permission to reprint must be obtained from the publisher. An earlier version by the author appeared on The Oil Drum website in August 2008, http://europe.theoildrum.com/node/4428.

Breaking the Fossil Fuel Habit

National governments everywhere seem to be in denial about peak oil. The citizens in at least one American state, though, have moved beyond denial and begun to prepare for life after peak oil.

In January of 2008, after a run-up of gasoline prices that spanned several years, Hawaii passed legislation to cut reliance on oil by 70 percent by 2030. Notably, this was done well before the big price spike in the summer of 2008, showing that the impetus was indeed more than economic cost. As Hawaiian environmental activist Jeffrey Mikulina put it, the state is one supertanker away from being Amish. And, with sea level rise, only years or decades away from being physically smaller. As for the ongoing politics of weaning the state off fossil fuels, some of the initiative is coming from a private entrepreneur, Henk Rogers and his Blue Planet Foundation. Rogers felt that the 2030 date was actually too far off so his foundation is trying to speed up the move toward energy independence by encouraging public and private sectors to act more quickly.

If Hawaii meets its goals, it may lead the way for other states and the country as a whole. After all, from a planetary perspective, we all live on a resource-constrained island.

Source: "Hawaii's Moon Shot," *New York Times*, December 2, 2008.

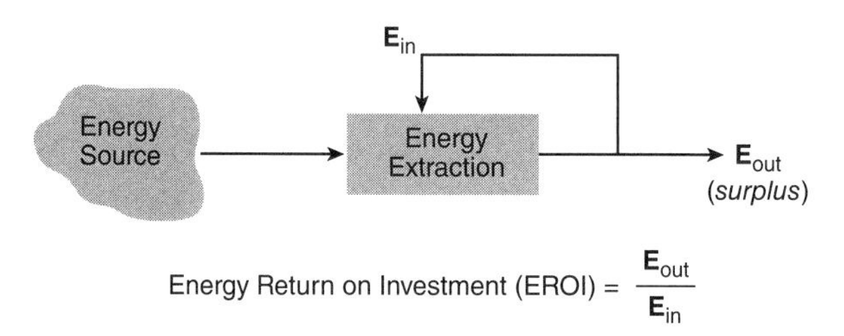

Figure 2.1
EROEI. *Source*: Charles Hall, Pradeep Tharakan, John Hallock, Wei Wu, and Jae-Young Ko, Advances in Energy Studies Conference, Porto Venere, Italy, September 2002, http://www.eroei.com/articles/the-chain/what-is-eroei/.

fuels for transport or electricity for lighting. It can also be energy consumed directly by living organisms—namely, food or space heat.

EROEI is mostly used by analysts and activists who focus on fossil fuel depletion issues. Cutler Cleveland, Charles Hall, and Robert Kaufmann, all energy policy analysts, Robert Costanza, an ecological economist, and Howard Odum, an ecologist, are leading proponents of the concept.[2]

However, within society's decision-making mainstream—financial markets, corporations, governments, parliaments, and their advisors in the civil service, policymaking arenas, and lobbying bodies—there is little evidence that the concept and significance of EROEI are grasped or accepted. Instead mainstream policymakers appraise different energy investment options by applying financial, political, and environmental criteria. Environmental criteria usually encompass climate change, local environmental effects, and waste management. Where resource constraints are discussed, the financial ROI is used, at least implicitly, as a proxy for EROEI. For example, the uranium reserve is normally described in terms of remaining tons of uranium that can be economically extracted at a minimum uranium commodity price ($US per kg)[3] using current technologies. Implicit in this formulation is the idea that financial cost and new technologies are the key determiners of availability, rather than any physical constraints. Nate Hagens, a former Wall Street hedge fund manager, has observed that the financial markets do not understand net energy.[4] Peter Davies, Special Economics Advisor at BP, the oil giant formerly called British Petroleum, has even gone so far as to say that the net energy of an energy source is an irrelevant criterion.[5]

In addition, mainstream decision makers tend to conduct energy policy on a predict-and-provide basis: the premise of energy policies is that "energy needs" must be met.[6] As a result, energy efficiency, defined as wealth per unit of energy input, has been encouraged since the oil crises of the 1970s. Indeed, industrialized economies now generate more economic wealth per unit of energy than ever before. *Total energy use*, however, has increased over time as the global economy has grown. Mainstream policymakers have not yet accepted planned reduction in total energy use as a matter of public policy, in part because they do not seem to understand the difference between increased energy efficiency and increased total use and in part because the impact of a real reduction would hamper future economic growth.[7] In the UK government's Department for Business Enterprise and Regulatory Reform's (BERR) May 2007 policy paper, "Meeting the Energy Challenge," there are fifteen references to the need to sustain "economic growth." No distinction is made between energy efficiency and total energy use.

Measuring EROEI—System Boundaries

While differences in philosophical outlook or political constraints may explain why EROEI has largely been ignored by mainstream decision makers, use of EROEI as a metric to appraise energy investment options, including total reduction options, is also problematic for practical reasons.

Currently no established, globally agreed-on criteria exist to define the boundaries of an energy system. What inputs should be counted as "energy invested"? At what point is the "energy returned" considered to have been delivered as a useful output? The results of an EROEI analysis and the conclusions that can be drawn from them are influenced strongly by the boundaries used to define an energy system.

Regarding "energy returned" and system boundaries, the scope of the energy investment option appraisal is crucial. If the scope of an EROEI analysis is limited to alternative methods of generating electricity for the national grid, the "energy returned" should be in the form of electricity delivered to the consumer, not that contained in the feedstock fuels (e.g., coal, gas, uranium). If the comparison is between methods of fueling vehicles, the energy return should be in the form of mechanical energy delivered to turn the vehicle's wheels, not the latent energy of the fuel itself.

Regarding energy invested, analysts must recognize that energy has to be invested at all stages in the life cycle of an energy system. As with financial accounting, some energy costs are directly associated with the activity of the system as a whole, not just its parts (e.g., the power plant or the transport vehicle). Other energy costs are overheads, which are allocated pro-rata. Some of the energy costs may have no connection to the energy system but nevertheless have been incurred elsewhere in the wider society. See table 2.1 for an example of a nearly complete listing of the energy costs of a nuclear reactor.

Even if a technologically simpler energy system were to be analyzed, such as a series of wind farms that generated the same net energy as the nuclear reactor, identifying and quantifying all the indirect costs (some of which are highly tenuous) would soon become impractical.

For the analyst and the policymaker, it is clearly more feasible to identify and quantify direct energy expenditures—for example, the energy costs of forging steel for a wind turbine tower, or the energy

Table 2.1
Energy costs of a nuclear reactor

Energy invested	Direct ←	COSTS	→	Indirect
Manufacture and installation	Embodied energy of components Transport of components	Share of energy needed to build factories Site security costs	Share of energy needed to support nuclear R&D programs (e.g., fast breeders, thorium, pebble reactors)	Share of energy cost for educating, feeding, and motivating staff involved with the construction
	Direct ←	COSTS	→	Indirect
Maintenance and fuel production during operational life	Energy used in mining, milling, enriching, fabricating, and transporting fuel Maintenance of plant Plant security	Share of energy needed for equipment, factories, and transport used in fuel production processes	Share of energy needed to support nuclear R&D programs (e.g., development of new methods of extracting from ever more marginal ores)	Energy cost of maintaining sufficiently stable international relations between region with the ore and region with the reactor
	Direct ←	COSTS	→	Indirect
Decommissioning/recycling of materials	Site clearance, shutting down and removal of equipment, sale of redundant assets Securing of radioactive materials	Cost of project management of decommissioning process: training, feeding, and motivating staff involved		
	Direct ←	COSTS	→	Indirect
Cleanup	Energy needed to build and maintain permanent facilities for long-term radioactive waste	Restoration of site to make it usable for other purposes after the reactor facility has been removed	Restoration of sites where ore was mined	

needed to transport the tower from the factory to its operational location. When does an energy cost become an indirect overhead? This might include a portion of the energy needed to build and run the foundry and its equipment, and perhaps a portion of the energy needed to train, feed, entertain, and motivate the foundry workers. A calculation including all these variables would rapidly become very time consuming and prone to error. Ultimately, a slice of the entirety of the remainder of society's activities could be apportioned to each energy-gathering activity, giving all energy sources an EROEI of 1:1.[8] Clearly an EROEI that included all indirect costs, however tenuous their association with the energy system, would not be a helpful tool for assessing how best to meet society's "energy needs." Despite these methodological issues relating to system boundaries, EROEI is still an essential tool in a peak oil era because it does what conventional measures (e.g., energy reserve and energy efficiency) do not do.

EROEI and Societal Complexity

So, if all the most tenuous, indirect energy costs of our global energy system were included in an EROEI analysis, the global energy system would have an EROEI of 1:1. This points to a more fundamental impact of EROEI on the nature of a society. All life expends energy in order to capture energy from its environment: if a fox does not obtain more energy from eating rabbits than it consumes catching them, it will not survive long; the same principle applies to all other life forms, from tulips to bacteria, whales, and humans. From the days of the earliest hunter-gatherers, the nature of human societies has been governed by their success in capturing energy (primarily food energy) at an energy profit.[9]

The question of which factors determine the fate of different societies throughout history has been addressed by Joseph Tainter in *The Collapse of Complex Societies*[10] and Jared Diamond in *Collapse*.[11] Both authors identify reasons why societies of all sizes, from small isolated settlements to the Roman Empire, have collapsed. Diamond identifies four reasons for collapse: resource depletion, climate change, hostile neighbors, and friendly neighbors. Tainter focuses on "energy gain" and complexity: societies that manage to achieve high energy gains from their energy systems develop complexity. Complex societies are characterized by high population densities, high levels of occupational specialization, steep social hierarchies, and high inequality.[12] As population densities and specialization increase, the energy systems that sustain them have to deliver more net energy. In doing so, they allow the population to grow and a greater percentage of that growing population becomes "nonproductive" specialists.

An EROEI analysis is consistent with Tainter's explanation of societal collapse and informs modern society's current predicament. Fossil fuels, which represent more than 80 percent of current total primary energy use,[13] have an EROEI that is declining,[14] eventually approaching the levels typical of current renewable energy technologies. If, as appears likely, modern industrial, high-consuming societies are indeed moving to a lower EROEI—that is, to lower-gain energy systems—the implications are profound.

Other Physical Criteria for Assessing the Practicability of Energy Systems

As we have seen, EROEI is a metric that can describe much more than the technical feasibility of an energy system. A comprehensive appraisal, however, would need several additional physical factors. Elements of these additional criteria could be incorporated into an EROEI analysis, while others fall outside it. Importantly, their incorporation makes all the more profound, possibly disturbing, the trends in declining EROEI.

1. *Infrastructural requirements.* The world is not a blank sheet in terms of energy systems; there is a lot of existing energy infrastructure. If a new energy system requires significant changes to the existing infrastructure, there will be a longer lead-in time, often measured in decades,[15] before the new technology starts to make a significant contribution to the energy mix. For example, if the current energy system delivers liquid fuel to feed millions of internal combustion engines, a new energy system that delivers electricity for electric vehicles will only be able to replace the old system as quickly as existing liquid-fuel vehicles can be replaced or retrofitted, a necessarily slow and costly process (financially and energetically). The energy costs of the infrastructural changes could be included as part of an EROEI analysis but the EROEI analysis would not be able to quantify the long lead-in times needed to adapt or replace energy infrastructure.

2. *Energy density.* This term denotes the ability to store energy where space and weight constraints apply. Most energy for transport, especially air and sea transport, needs an onboard energy store. The new energy

system needs to produce energy in a form that can be stored with existing technology. The storage technology will probably require conversion of the energy, with consequent losses. This criterion could be taken into account in an EROEI analysis, if the storage process is included within the EROEI system boundary.

3. *Location of resource (in relation to demand).* If the energy source is far from where it is needed, the source, even if it has a high EROEI, may be impractical. This criterion could be incorporated in an EROEI analysis, if the transport process and its energy costs are included within the system boundary.

4. *Scalability and rate of extraction.* The rate at which the resource (whether renewable or finite) can be harvested will determine the maximum flow rate of energy that the energy system can deliver—for instance, barrels per day, kW of electricity. Similarly, the size of the resource, together with the maximum flow rate, will determine how big its contribution to the energy system can be and, if the source is nonrenewable, for how long.

5. *Environmental impact.* The extent to which an energy system affects the global and local environment has to be taken into account. This is seen most clearly with fossil fuels and the effect they are having on the global climate. An EROEI analysis would not normally incorporate this factor, unless the energy costs of correcting the environmental impact were incorporated under energy invested. Here is where climate regulation and ecosystem functioning would be a useful complement to EROEI. The link between declining EROEI and environmental impacts has been explored in more detail by Williams, Warr, and Ayres (2008).[16]

6. *Complexity and resilience.* As discussed, there is a direct link between the EROEI of society's energy systems and the complexity of the society that can be sustained by it. Political scientist Thomas Homer-Dixon[17] links Joseph Tainter's work to the work of Crawford S. Holling on ecological systems and concludes that all complex systems, including those that make up and support human society, go through "adaptive cycles" of growth, collapse, and regeneration. As a system grows, it not only becomes more complex, it also develops greater connectedness, as each part becomes dependent on the part before it. As human actors in the system seek ever-greater EROEI to support their increasingly complex society, they use their ingenuity to improve the efficiency of each part of the system. Doing so makes the relationship between the parts more tightly coupled and less adaptive. The efficiency gain is at the expense

of system resilience to unexpected shocks. As the system loses resilience, it becomes "brittle"—that is, less able to cope with relatively short interruptions in the flow of system inputs. Comparing the current, modern, globalized economic system, using just-in-time logistics facilitated by IT and the Internet, with its 1970s counterparts illustrates how much less resilient our support systems are today. It appears that, at least anecdotally, there are far more potential failure points today than thirty years ago.

In making choices about energy systems, if a complex and brittle one is chosen (however inadvertently) over a less complex and more resilient one, a failure in one part is likely to become a failure across the wider system. A complex and brittle energy system itself requires high levels of complexity in the wider society to be maintained. Such complexity could be risky when a society's main sources of energy, fossil fuels, are finite and expected to go into decline during the next two or three decades and where the timing of some of the declines, particularly coal,[18] is highly uncertain.

7. *Managing supply and demand (storage issues).* Energy provision could be compared with performing a play in a theater—one that never ends! A show that must go on: day after day, year after year, following demand. The issue of predict-and-provide needs to be examined carefully and critically when appraising options for electricity generation systems, particularly options that fluctuate and are geographically distributed such as wind, wave, tides, or solar. Many countries will need to modify and augment their electricity grids, as well as increase storage capacity, to accommodate high percentages of these generation sources. The energy cost of some of this additional infrastructural investment should be included in an EROEI analysis of a renewables-rich electricity supply. These issues, as they affect the United Kingdom, are explored in more detail by Boyle and colleagues[19] and in the *Zero Carbon Britain 2030* report,[20] and they can be applied to other countries with electricity grids.

Minimum EROEI Needed to Support a Complex Society

The question of what minimum EROEI is needed to support a complex society has been considered by Hall, Balogh, and Murphy.[21] Figure 2.2 illustrates graphically how dramatically energy use increases as a proportion of all human activity, once EROEI drops into single figures.

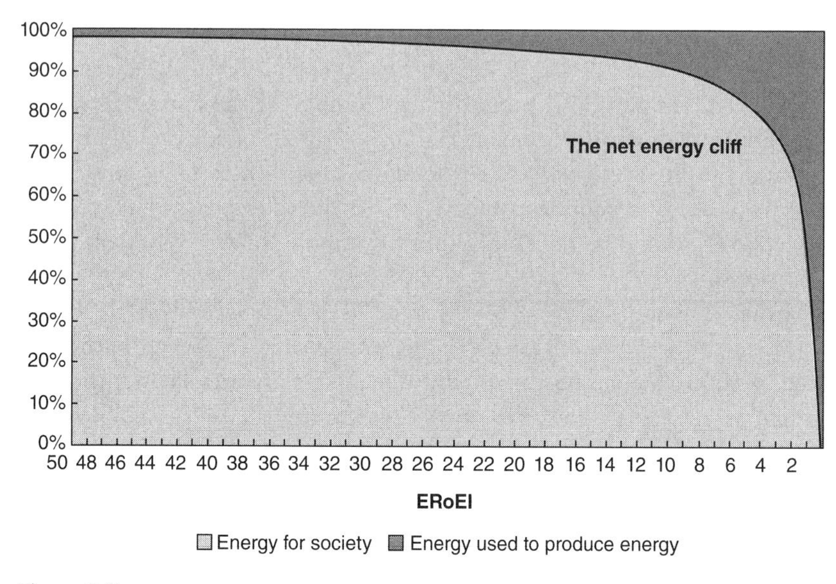

Figure 2.2
Net energy cliff. *Source*: Euan Mearns, *ASPO-USA: Support for Global Energy Flow Modelling and a Net Energy Database*, The Oil Drum, 2006, http://europe.theoildrum.com/story/2006/10/31/144929/65.

The concept of a minimum EROEI is useful in assessing so-called nonproductive human activities, activities that are seen as desirable or useful or essential and yet do not have a high return on energy invested; healthcare is a case in point. But the concept also helps decision makers build in a buffer against unexpected, adverse environmental changes, changes that could reduce the ability of a society to extract useful net energy. Two nontrivial examples are the effects of global climate destabilization on current and future agricultural production; and, rather ironically, fossil fuel extraction in the Gulf of Mexico that is increasingly endangered by the growing frequency and severity of tropical storms, a likely consequence of burning fossil fuels.

Hall et al. (2009) analyze some of the indirect costs—transport, refining, infrastructure—incurred in providing and using fossil fuel energy services to the consumer to arrive at an "extended EROEI" ($EROEI_{ext}$) figure, as distinct from the well- or minehead ($EROEI_{mm}$), which considers only direct costs. They estimate that an extended $EROEI_{ext}$ will be approximately *a third* of the wellhead $EROEI_{mm}$. That is, once all measurable energy costs are taken into account, actual available energy may be only a third of current estimates.

Current global $EROEI_{mm}$ is estimated at 20:1, down from 35:1 a decade ago. Applying Hall's method, the current global $EROEI_{ext}$ is actually about 6:1, right at the inflection point in figure 2.2 where society has to start directing much bigger proportions of its activity to energy production. Hall considers an $EROEI_{ext}$ of 3:1 to be the "bare minimum" for civilization in that it would offer society "little discretionary surplus" to support such activities as "art, medicine, education and so on." If, as expected, the downward trend of global EROEI continues, globally we could reach this bare minimum by the end of the 2020s, although there could be significant regional variations.

Conclusion

EROEI is, or should be, the most important physical criterion used to assess the practicability of new energy systems for two reasons. First, if the EROEI of an energy system is 1:1 or lower, it is no longer an energy source. As the EROEI drops below 1:1, it becomes an energy sink. That is, if it takes more energy to extract than one gains by that extraction, then one is throwing away usable energy; there is a net loss. This is important now, because society currently benefits hugely from the very high EROEIs of fossil fuels. From the fossil fuel perspective it is hard to decipher, psychologically and institutionally, the implications of switching to the unavoidably low EROEI alternatives such as biofuels and solar. The growth in use of corn-based ethanol as a substitute for fossil fuel in the U.S. vehicle fleet is one example: a biofuel-based energy system may deliver no net energy.[22]

Second, the energy choices made now, if they are not made with a grasp of the wider implications of a reduction in our energy systems' overall EROEI and overall reduction in total energy use, will cause a profound, painful, and largely unexpected and apparently inexplicable reduction in the complexity of society: in other words, an unmanaged and protracted collapse.

Despite the centrality of EROEI, the other physical criteria referred to above must also be taken into account in an appraisal of energy-system options, although energy density, location of resource, and (to a certain extent) infrastructural requirements could all be incorporated into an EROEI analysis. I have assumed that the analyses would compare different energy supply alternatives. An EROEI analysis could equally usefully be applied to different conservation and demand reduction alternatives. In these cases, the other physical criteria may be different. Appraising the effect of psychological factors and behavior change would require a different analytical approach.

Heinberg has stressed the importance of making our energy choices "carefully, intelligently and co-operatively."[23] Clearly nonphysical factors also influence an energy-system option appraisal: financing/capital constraints (particularly in the new financial reality post the crash of 2008), psychological and behavioral factors, political and security issues, as well as macroeconomic implications. Although these remain important, if our energy choices are going to be careful and intelligent, the physical criteria, particularly EROEI, must be fulfilled, or at least the extent to which they act as constraints must be understood, before the nonphysical factors are considered. This means most importantly that the nonphysical factors must not be seen as key while the physical criteria are dismissed as peripheral or irrelevant, or treated as if market pricing and new technologies can always overcome physical constraints. This overview of the physical criteria that determine a proposed energy system's usefulness in an era of declining fossil fuel availability and climate disruption illustrates why EROEI is the most important but not the sole criterion in deciding how to meet society's energy needs.

Notes

1. C. J. Cleveland, "Net Energy from the Extraction of Oil and Gas in the United States," *Energy* 30, no. 5 (2005): 769–782, Elsevier, http://tinyurl.com/3vuplz6.

2. C. J. Cleveland, R. Costanza, C. A. S. Hall, and R. Kaufmann, "Energy and the US Economy: A Biophysical Perspective," *Science* 225, no. 4665 (1984): 890–897; H. T. Odum, *Environmental Accounting: Emergy and Environmental Decision Making* (New York: Wiley, 1996).

3. From http://www.world-nuclear.org/info/inf75.html.

4. In a radio interview with Jason Bradford on KZYX on June 5, 2006; 8 minutes and 35 seconds into the audio archive at http://old.globalpublicmedia.com/nate_hagens_on_the_partys_over_going_local.

5. Presentation by Peter Davies, BP Special Economic Advisor, to the All Party Parliamentary Group on Peak Oil and Gas, January 16, 2008, http://www.appgopo.org.uk/index.php?option=com_content&task=view&id=21 and http://www.appgopo.org.uk/images/documents/bp_appgopo_16_jan_2008.pdf.

6. Secretary of State for Trade & Industry, *Meeting the Energy Challenge*, UK Government White Paper on Energy, Department for Business Enterprise and Regulatory Reform (BERR), formerly the Department for Trade and Industry (DTI), CM7124/ID5539714 (London: The Stationery Office/HMSO, May 2007), http://www.berr.gov.uk/files/file39387.pdf.

7. D. I. Stern, *Energy & Economic Growth*, Rensselaer Polytechnic Institute, Troy, NY, April 2003, http://www.localenergy.org/pdfs/Document%20Library/Stern%20Energy%20and%20Economic%20Growth.pdf.

8. J. Vail, *EROEI Short #2: Lenin & Lohan*, The Oil Drum, August 28, 2007, http://www.theoildrum.com/node/2893.

9. R. Heinberg, speaking at the Annual Conference of the Soil Association in Cardiff, January 26, 2007, summarized by R. Hopkins and P. Holden, *One Planet Agriculture: Handbook for Practical Action* (Bristol, UK: Soil Association, 2007), http://www.soilassociation.org/LinkClick.aspx?fileticket=s2bD6uoUYqo%3d&tabid=387.

10. J. A. Tainter, *The Collapse of Complex Societies* (Cambridge: Cambridge University Press, 1988) (paperback ed., 1990).

11. J. Diamond, *Collapse: How Societies Choose to Fail or Survive* (London: Penguin, 2005).

12. I. Illich, *Energy and Equity* (London: Calder and Boyars, 1974).

13. International Energy Agency, *World Energy Outlook 2006* (Paris: OECD/IEA, 2006), 492 (p. 493 of the PDF file), http://www.iea.org/textbase/nppdf/free/2006/weo2006.pdf.

14. C. J. Cutler, "Net Energy from the Extraction of Oil and Gas in the United States," *Energy* 30 (2005): 769–782, www.sciencedirect.com / www.elsevier.com/locate/energy.

15. Robert L. Hirsch Roger Bezdek, and Robert Wendling, *Peaking of World Oil Production: Impacts, Mitigation, & Risk Management* (Washington, DC: U.S. Department of Energy, February 2005), http://www.netl.doe.gov/publications/others/pdf/Oil_Peaking_NETL.pdf.

16. E. Williams, B. Warr, and R. U. Ayres, "Efficiency Dilution: Long-Term Exergy Conversion Trends in Japan," *Environmental Science & Technology* 42, no. 13 (2008): 4964–4970, http://pubs.acs.org/doi/abs/10.1021/es071719a.

17. T. F. Homer-Dixon, *The Upside of Down: Catastrophe, Creativity, and the Renewal of Civilization* (Washington, DC: Island Press, 2006), http://www.theupsideofdown.com/.

18. Werner Zittel and Jörg Schindler, *Coal: Resources and Future Production*, Energy Watch Group, EWG Paper No. 1/07, March 2007, updated version dated July 10, 2007, http://www.energywatchgroup.org/fileadmin/global/pdf/EWG_Report_Coal_10-07-2007ms.pdf.

19. G. Boyle, ed., *Renewable Electricity and the Grid: The Challenge of Variability* (London: Earthscan, 2007).

20. M. Kemp, ed., *Zero Carbon Britain 2030* (Machynlleth, UK: CAT Publications, 2010), http://www.zerocarbonbritain.org.

21. C. A. S. Hall, S. Balogh, and D. J. R. Murphy, "What Is the Minimum EROI That a Sustainable Society Must Have?," *Energies* 2 (2009): 25–47, http://www.mdpi.com/1996-1073/2/1/25/pdf.

22. C. J. Cleveland et al., "Energy Returns on Ethanol Production," *Science*, June 23, 2006, 1746–1748, http://www.sciencemag.org/cgi/content/full/312/5781/1746.

23. R. Heinberg, "Peak Oil: Local Solutions to a Global Challenge," lecture from *Life After Oil* course at Schumacher College, Dartington, Totnes, Devon, UK, November 2006, delivered at Totnes Civic Hall Transition Town, November 22, 2006; recordings available on DVD from Schumacher College at http://www.schumachercollege.org.uk/news/culture-community-and-home-open-evening-report.

3

The Inevitability of Transition*

Joseph A. Tainter

Americans have long believed that their society is exceptional, its technological prowess rendering inconsequential the biophysical limits that destabilized past civilizations. Anthropologist and historian Joseph A. Tainter counters this belief by analyzing eighteen distinctive civilizations that each reached growth limits and then collapsed. Tainter explains these civilizations' descent with a straightforward application of the law of diminishing marginal returns—in this case, returns on increasing social complexity.

Two nuances in his work are especially pertinent to localization. First, contrary to what the term implies, collapse in this context does not mean total destruction but rather a rapid descent to a lower level of societal complexity. Second, technological innovation cannot prevent this descent; rather, technology, in its many forms, is the means by which society solves emergent problems. In so doing, society increases its complexity, eventually encountering diminishing marginal returns and thus increased vulnerability to breakdown.

Tainter's work has a sobering implication: while it may make sense to intentionally simplify one's own society to adapt to biophysical limits, in a world of globally interlocked economies other societies must do likewise; if they do not, mutual descent is likely. Therefore, a strong and meaningful notion of localization must go beyond the local. It must be at once local, regional, and international in scope. Isolationist strategies at any level are unlikely to succeed.

Collapse is recurrent in human history; it is global in its occurrence; and it affects the spectrum of societies from simple foragers to great empires.

*Tainter, Joseph A. (1988). "Summary and implications." In *The Collapse of Complex Societies*, 193–215. Cambridge: Cambridge University Press. Excerpted and reprinted with permission.

Collapse is a matter of considerable importance to every member of a complex society, and seems to be of particular interest to many people today. Political decentralization has repercussions in economics, art, literature, and other cultural phenomena, but these are not its essence. Collapse is fundamentally a sudden, pronounced loss of an established level of sociopolitical complexity.

A complex society that has collapsed is suddenly smaller, simpler, less stratified, and less socially differentiated. Specialization decreases and there is less centralized control. The flow of information drops, people trade and interact less, and there is overall lower coordination among individuals and groups. Economic activity drops to a commensurate level, while the arts and literature experience such a quantitative decline that a dark age often ensues. Population levels tend to drop, and for those who are left the known world shrinks.

Complex societies, such as states, are not a discrete stage in cultural evolution. Each society represents a point along a continuum from least to most complex. Complex forms of human organization have emerged comparatively recently, and are an anomaly of history. Complexity and stratification are oddities when viewed from the full perspective of our history, and where present, must be constantly reinforced. Leaders, parties, and governments need constantly to establish and maintain legitimacy. This effort must have a genuine material basis, which means that some level of responsiveness to a support population is necessary. Maintenance of legitimacy or investment in coercion requires constant mobilization of resources. This is an unrelenting cost that any complex society must bear.

Two major approaches to understanding the origin of the state are the conflict and integration schools. The former sees society as an arena of class conflict. The governing institutions of the state, in this view, arose out of economic stratification, from the need to protect the interests of propertied classes. Integration theory suggests, in contrast, that governing institutions (and other elements of complexity) emerged out of society-wide needs, in situations where it was necessary to centralize, coordinate, and direct disparate subgroups. Complexity, in this view, emerged as a process of adaptation.

Both approaches have strong and weak points, and a synthesis of the two seems ultimately desirable. Integration theory is better able to account for distribution of the necessities of life, conflict theory for surpluses. There are definitely beneficial integrative advantages in the concentration of power and authority, but once established the political realm becomes an increasingly powerful influence. In both views, though, the state is a problem-solving organization, emerging because of changed circumstances (differential economic success in the conflict view; management of society-wide stresses in integration theory). In both approaches legitimacy, and the resource mobilization this requires, are constant needs. . . .

Four concepts lead to understanding collapse, the first three of which are the underpinnings of the fourth. These are:

1. human societies are problem-solving organizations;
2. sociopolitical systems require energy for their maintenance;
3. increased complexity carries with it increased costs per capita; and
4. investment in sociopolitical complexity as a problem-solving response often reaches a point of declining marginal returns.

This process has been illustrated for recent history in such areas as agriculture and resource production, information processing, sociopolitical control and specialization, and overall economic productivity. In each of these spheres it has been shown that industrial societies are experiencing declining marginal returns for increased expenditures. The reasons for this are summarized below.

To the extent that information allows, rationally acting human populations first make use of sources of nutrition, energy, and raw materials that are easiest to acquire, extract, process, and distribute. When such resources are no longer sufficient, exploitation shifts to ones that are costlier to acquire, extract, process, and distribute, while yielding no higher returns.

Information processing costs tend to increase over time as a more complex society requires ever more specialized, highly trained personnel, who must be educated at greater cost. Since the benefits of specialized training are always in part attributable to the generalized training that must precede it, more technical instruction will automatically yield a declining marginal return. Research and development move from generalized knowledge that is widely applicable and obtained at little cost, to specialized topics that are more narrowly useful, are more difficult to resolve, and are resolved only at great cost. Modern medicine presents a clear example of this problem.

Sociopolitical organizations constantly encounter problems that require increased investment merely to preserve the status quo. This investment comes in such forms as increasing size of bureaucracies, increasing specialization of bureaucracies, cumulative organizational solutions, increasing costs of legitimizing activities, and increasing costs of internal control and external defense. All of these must be borne by levying greater costs on the support population, often to no increased advantage. As the number and costliness of organizational investments increase, the proportion of a society's budget available for investment in future economic growth must decline.

Thus, while initial investment by a society in growing complexity may be a rational solution to perceived needs, that happy state of affairs cannot last. As the least costly extractive, economic, information-processing, and organizational solutions are progressively exhausted, any further need for increased complexity must be met by more costly responses. As the cost of organizational solutions grows, the point is reached at which continued investment in complexity does not give a proportionate yield, and the marginal return begins to decline. The added benefits per unit of investment start to drop. Ever greater increments of investment yield ever smaller increments of return.

A society that has reached this point cannot simply rest on its accomplishments, that is, attempt to maintain its marginal return at the status quo, without further deterioration. Complexity is a problem-solving strategy. The problems with which the universe can confront any society are, for practical purposes, infinite in number and endless in variety. As stresses necessarily arise, new organizational and economic solutions must be developed, typically at increasing cost and declining marginal return. The marginal return on investment in complexity accordingly deteriorates, at first gradually, then with accelerated force. At this point, a complex society reaches the phase where it becomes increasingly vulnerable to collapse.

Two general factors can make such a society liable to collapse. First, as the marginal return on investment in complexity declines, a society invests ever more heavily in a strategy that yields proportionately less. Excess productive capacity and accumulated surpluses may be allocated to current operating needs. When major stress surges (major adversities) arise there is little or no reserve with which they may be countered. Stress surges must be dealt with out of the current operating budget. This often proves ineffectual. Where it does not, the society may be economically weakened and made more vulnerable to the next crisis.

Once a complex society enters the stage of declining marginal returns, collapse becomes a mathematical likelihood, requiring little more than sufficient passage of time to make probable an insurmountable calamity. So if Rome had not been toppled by Germanic tribes, it would have been later by Arabs or Mongols or Turks. A calamity that proves disastrous to an older, established society might have been survivable when the marginal return on investment in complexity was growing. Rome, again an excellent example, was thus able to withstand major military disasters during the Hannibalic war (late third century B.C.), but was grievously weakened by losses that were comparatively less (in regard to the size and wealth of the Roman state at these respective times) at the Battle of Hadrianople in 378 A.D. Similarly, the disastrous barbarian invasions of the first decade of the fifth century were actually smaller than those defeated by Claudius and Probus in the late third century (Dill 1899, 299).

Secondly, declining marginal returns make complexity an overall less attractive strategy, so that parts of a society perceive increasing advantage to a policy of separation or disintegration. When the marginal cost of investment in complexity becomes noticeably too high, various segments increase passive or active resistance, or overtly attempt to break away. The insurrections of the Bagaudae in late Roman Gaul are a case in point.

At some point along the declining portion of a marginal return curve, a society reaches a state where the benefits available for a level of investment are no higher than those available for some lower level. . . . Complexity at such a point is decidedly disadvantageous, and the society is in serious danger of collapse from decomposition or external threat.

. . . [Four] major topics remain to be addressed. These are: (1) further observations on collapse, and on the nature of the declining productivity of complexity; (2) application and extension of the concept; . . . [3] subsuming other explanatory themes under declining marginal returns; and [4] implications for contemporary times and for the future of industrial societies. . . . The definition of collapse will be completed here.

Collapse and the Declining Productivity of Complexity

We arrive in this section at one of the major implications of the study. Most of the writers whose work has been considered seem to approve of civilizations and complex societies. They see complexity as a desirable, even commendable, condition of human affairs. Civilization to them is the ultimate accomplishment of human society, far preferable to simpler, less differentiated forms of organization. An appreciation for the artistic, literary, and scientific accomplishments of civilizations clearly has much to do with this, as does the industrial world's view of itself as the culmination of human history. Toynbee is perhaps most extreme in this regard, but he is by no means atypical. Spengler, in his abhorrence of civilization and its sequelae, represents a minority view, as does Rappaport.

With such emphasis on civil society as desirable, it is almost necessary that collapse be viewed as a catastrophe. An end to the artistic and literary features of civilization, and to the umbrella of service and protection that an administration provides, are seen as fearful events, truly paradise lost. The notion that collapse is a catastrophe is rampant, not only among the public, but also throughout the scholarly professions that study it. Archaeology is as clearly implicated in this as is any other field. As a profession we have tended disproportionately to investigate urban and administrative centers, where the richest archaeological remains are commonly found. When with collapse these centers are abandoned or reduced in scale, their loss is catastrophic for our data base, our museum collections, even for our ability to secure financial backing. (Dark ages are rarely as attractive to philanthropists or funding institutions.) Archaeologists, though, are not solely at fault. Classicists and historians who rely on literary sources are also biased against dark ages, for in such times their data bases largely disappear.

A less biased approach must be not only to study elites and their creations, but also to acquire information on the producing segments of complex societies that continue, if in reduced numbers, after collapse. Archaeology, of course, has great potential to provide such information.

Complex societies, it must be emphasized again, are recent in human history. Collapse then is not a fall to some primordial chaos, but a return to the normal human condition of lower complexity. The notion that collapse is uniformly a catastrophe is contradicted, moreover, by the present theory. To the extent that collapse is due to declining marginal returns on investment in complexity, it is an *economizing* process. It occurs when it becomes necessary to restore the marginal return on organizational investment to a more favorable level. To a population that is receiving little return on the cost of supporting complexity, the loss of that complexity brings economic, and perhaps administrative, gains. Again, one is reminded of the support sometimes given by the later Roman population to the invading barbarians, and of the success of the latter at deflecting further invasions of western Europe. The attitudes of the late Maya and Chacoan populations toward their administrators cannot be known, but can easily be imagined.

Societies collapse when stress requires some organizational change. In a situation where the marginal utility of still greater complexity would be too low, collapse is an economical alternative. Thus the Chacoans did not rise to the challenge of the final drought because the cost of doing so would have been too high relative to the benefits. Although the end of the Chacoan system meant the end of some benefits (as does the end of any complex system), it also brought an increase in the marginal return on organization. The Maya, similarly, appear to have reached the point where evolution toward larger polities would have brought little return for great effort. Since the status quo was so deleterious, collapse was the most logical adjustment.

One of the explanatory themes . . . the "failure to adapt" model—may now have its full weakness revealed. Proponents of this view argue, in one form or another, that complex societies end because they fail to respond to changed circumstances. This notion is clearly obviated: *under a situation of declining marginal returns collapse may be the most appropriate response*. Such societies have not failed to adapt. In an economic sense they have adapted well—perhaps not as those who value civilizations would wish, but appropriately under the circumstances.

What may be a catastrophe to administrators (and later observers) need not be to the bulk of the population (as discussed, for example, by Pfeiffer [1977: 469–71]). It may only be among those members of a society who have neither the opportunity nor the ability to produce primary food resources that the collapse of administrative hierarchies is a clear disaster. Among those less specialized, severing the ties that link local groups to a regional entity is often attractive. Collapse then is not intrinsically a catastrophe. It is a rational, economizing process that may well benefit much of the population. . . .

Further Implications of Declining Marginal Returns

It may seem from this work that archaeology is campaigning to displace economics as the "dismal science." Of course, the marginal product curve is nothing new. It was developed to characterize changing cost/benefit curves in resource extraction, and input/output ratios in the manufacturing sector. The idea of diminishing returns to economic activity is at least as old as the nineteenth-century classical economists: Thomas Malthus, David Ricardo, and John Stuart Mill (Barnett and Morse 1963, 2). It applies . . . to subsistence agriculture, minerals and energy production, information processing, and to many features of sociopolitical organization. Wittfogel (1955, 1957) applied the concept of "administrative returns" to the extension of government into economic affairs in "Oriental Despotisms." Lattimore (1940) accounted for the Chinese dynastic cycle in terms of increasing and declining returns. It seems that Kroeber's (1957) observations on the "fulfillment" of art styles may refer to a situation where innovation within a style becomes increasingly difficult to achieve, leading to repetition and rearrangement of earlier work, and ultimately to a new style in which innovation is more easily attained. The phenomenon is not at all limited to the human species. Animal predators seem to follow the principle of marginal returns in their selection of environmental patches in which to forage (Charnov 1976; Krebs 1978, 45–48). . . .

There are significant differences in the evolutionary histories of societies that have emerged as isolated, dominant states, and those that have developed as interacting sets of what Renfrew (1982, 286–289) has called "peer polities" and B. Price has labeled "clusters" (1977). Renfrew's term is appropriately descriptive. Peer polities are those like the Mycenaean states, the later small city-states of the Aegean and the Cyclades, or the centers of the Maya Lowlands, that interact on an approximately equal level. As Renfrew and Price make clear, the evolution of such clusters of peer polities is conditioned not by some dominant neighbor, but more usually by their own mutual interaction, which may include both exchange and conflict.

In competitive, or potentially competitive, peer polity situations the option to collapse to a lower level of complexity is an invitation to be dominated by some other member of the cluster. To the extent that such domination is to be avoided, investment in organizational complexity must be maintained at a level comparable to one's competitors, *even if marginal returns become unfavorable*. Complexity must be maintained regardless of cost. Such a situation seems to have characterized the Maya, whose individual states developed as peer polities for centuries, and then collapsed within a few decades of each other (Sabloff 1986).

The post-Roman states of Europe have experienced an analogous situation, especially since the demise of the Carolingian Empire. European history of the past 1500 years is quintessentially one of peer polities interacting and competing, endlessly jockeying for advantage, and striving to either expand at a neighbor's expense or avoid having the neighbor do likewise. Collapse is simply not possible in such a situation unless all members of the cluster collapse at once. Barring this, any failure of a single polity will simply lead to expansion of another, so that no loss of complexity results. The costs of such a competitive system, as among the Maya, must be met by each polity, however unfavorable the marginal return. . . .

At this point we arrive at the first step toward understanding the difference between societies that slowly disintegrate and those that rapidly collapse. The Byzantine and Ottoman empires are classic examples of the former. Both gradually lost power and territory to competitors. There was in this process no collapse—no sudden loss of complexity—for each episode of weakness by these empires was simply met by expansion of their neighbors. Herein lies an important principle of collapse. . . . *Collapse occurs, and can only occur, in a power vacuum.* Collapse is possible only where there is no competitor strong enough to fill the political vacuum of disintegration. Where such a competitor does exist there can be no collapse, for the competitor will expand territorially to administer the population left leaderless. Collapse is not the same thing as change of regime. Where peer polities interact collapse will affect all equally, if and when it occurs, provided that no outside competitor is powerful enough to absorb all.

Here, then, is the reason why the Mayan and Mycenaean centers collapsed simultaneously. No mysterious invaders captured each of these polities in an improbable series of fairy-tale victories. As the Mayan and Mycenaean petty states became respectively locked into competitive spirals, each had to make ever greater investments in military strength and organizational complexity. As the marginal return on these investments declined, no polity had the option to simply withdraw from the spiral, for this would have led to absorption by a neighbor. Collapse for such clusters of peer polities must be essentially simultaneous, as together they reach the point of economic exhaustion. Since in both cases no outside dominant power (in the Mesoamerican Highlands or the eastern Mediterranean) was both close enough and strong enough to take advantage of this exhaustion, collapse proceeded without external interference and lasted for centuries. . . .

Declining Marginal Returns and Other Theories of Collapse

The extent to which a global theory is illuminating or trivial depends, in part, on its ability to clarify matters that were previously obscure, on its flexibility in application, and on its power to incorporate less general explanations. The perspective of declining marginal returns has indeed clarified the collapse process and shown itself highly flexible in application. . . .

Resource depletion. The essence of depletion arguments is the gradual or rapid loss of at least part of a necessary resource base, whether due to agricultural mismanagement, environmental fluctuation, or loss of trade networks. Major weaknesses of the approach are: why steps are not taken to halt the approaching weakness; and why resource stress leads to collapse in one case and economic intensification in another. Consideration here must be given to the cost of further economic intensification projected against the marginal benefits to be gained. If the marginal utility of further economic development is too low, and/or if a society is already economically weakened by a low marginal return, then collapse in such instances would be understandable. Collapse is not understandable, under resource stress, without reference to characteristics of the society, most particularly its position on a marginal return curve. A society already experiencing a declining marginal return may not be able to capitalize the economic development that is often a response to resource stress.

New resources. The most general statement of this theme has been given by Harner (1970), who argues that new resources can alleviate shortages and inequities, ending the need for ranking and complexity. This can be squarely subsumed under declining marginal returns: when a system of ranking and complexity is no longer needed, continued support of it would yield a declining return, and so it is likely to be dropped.

Catastrophes. Catastrophe theories suffer from the same flaw as resource depletion arguments. Why, when complex social systems are designed to handle catastrophes and routinely do, would any society succumb? If any society has ever succumbed to a single-event catastrophe, it must have been a disaster of truly colossal magnitude. Otherwise, the inability of a society to recover from perturbation must be attributable to economic weakness, resulting quite plausibly from declining marginal returns.

Insufficient response to circumstances. The "failure to adapt" model relies on a value judgement: that complex societies are preferable to simpler ones, so that their disappearance must indicate an insufficient response. It ignores the possibility that, due to declining marginal returns, collapse may be an economical and highly appropriate adjustment. One major theory under this theme, Service's "Law of Evolutionary Potential," [is] subsumable under the principle of declining marginal returns. Conrad and Demarest's (1984) study shows how the Aztec and Inca empires reached the point of diminishing return for expansion, and declined accordingly. Other theories grouped under this theme are not plausibly linked to collapse.

Other complex societies. Blanton's argument that Monte Alban collapsed when it was no longer necessary for some tasks (deterrence of Teotihuacan) nor efficient at others (adjudication of disputes), is fully compatible with the marginal return principle. Monte Alban collapsed, in other words, when the return it could offer became too low relative to support costs. In regard to inter-polity competition, John Hicks once suggested that ". . . when the ability to expand is lost, the ability to recover from disasters may go too" (1969, 59). The ability to expand may be lost due to an economic weakness, or else where the cost of expansion becomes too high relative to advantages. The latter will occur where one complex society impinges on another (e.g., Rome and Persia), and the marginal return for conquest and administration is too low.

Intruders. The scenario of tribal peoples toppling great empires presents a major explanatory puzzle. What characteristics of the less complex society and/or what weaknesses in the more complex one could lead to such a circumstance? Service, as noted, ascribed this to his Law of Evolutionary Potential, which as pointed out can be subsumed under the principle of declining marginal returns. As discussed in regard to the ideas of Polybius and Service, a more powerful state may not prevail against a weaker one if the latter is ascending a marginal return curve and the former descending. A complex society that is investing heavily in many cumulative organizational features, with low marginal return, may have little or no reserve for containing stress surges. Such a state may compete inefficiently with a population that is smaller, and on paper weaker, but that invests in little but high-return military ventures.

Conflict/contradictions/mismanagement. . . . Class conflict is more likely a matter of a falling than a rising marginal return. In a case where the marginal return is rising, class conflict may be forestalled by creating the impression that opportunity for improvement exists for all classes. . . .

Social dysfunction. This vague theme is somewhat diverse, but its central concern seems to be with mysterious internal processes that prevent either integration or proper adaptation. Little understanding is gained by such ethereal notions. Much more would be learned by focusing on the costs and benefits of adopting complex social features.

Mystical. The mystical theme is difficult to incorporate under any scientific approach, but some of the individual studies grouped under this theme can be subsumed under the principle of declining marginal returns. David Stuart, for example, asserts that complex societies experience cyclical oscillations between more and less complex forms (which he labels "powerful" and "efficient"). The mystical nature of Stuart's formulation emerges when he cannot account for these oscillations, except to liken complex societies to insect swarms and to suggest that they "burn out" (Stuart and Gauthier 1981, 10–11). Why do Stuart's "powerful" societies revert to "efficient" ones? The answer is most likely that they do so because, as complex societies, they experience a declining marginal return on investment in complexity, and so become liable to collapse. . . .

Chance concatenation of events. Chance concatenations cannot explain collapse, except where combinations of deleterious circumstances impinge on a society already economically weakened.

Economic explanations. The themes that unite economic explanations are declining advantages to complexity, increasing disadvantages to complexity, and/or increasing costliness of complexity. Such ideas are clearly subsumable under declining marginal returns, and indeed this principle provides the global applicability previously lacking in economic explanations. . . .

The principle of declining marginal returns is indeed, then, capable of incorporating these various approaches to collapse (or at least the more worthwhile parts of these). It provides an overarching theoretical framework that unites diverse approaches, and it shows where connections exist among disparate views. It seems from this discussion that a significant range of human behaviors, and a number of social theories, are clarified by this principle.

Contemporary Conditions

A study of this topic must at some point discuss implications for contemporary societies, not only as a matter of social responsibility, but also because the findings point so clearly in that direction. Complex societies historically are vulnerable to collapse, and this fact alone is disturbing to many. Although collapse is an economic adjustment, it can nevertheless be devastating where much of the population does not have the opportunity or the ability to produce primary food resources.

Many contemporary societies, particularly those that are highly industrialized, obviously fall into this class. Collapse for such societies would almost certainly entail vast disruptions and overwhelming loss of life, not to mention a significantly lower standard of living for the survivors. . . .

. . . Patterns of declining marginal returns can be observed in at least some contemporary industrial societies in the following areas:

agriculture;
minerals and energy production;
research and development;
investment in health;
education;
government, military, and industrial management;
productivity of GNP for producing new growth; and
some elements of improved technical design.

. . . There are two opposing reactions to such trends. On the one hand there are a number of economists who, despite the reputation of their discipline for pessimism, believe that we face, not real resource shortages, only solvable economic dilemmas. They assume that with enough economic motivation, human ingenuity can overcome all obstacles. Three quotations characterize this approach:

No society can escape the general limits of its resources, but no innovative society need accept Malthusian diminishing returns (Barnett and Morse 1963, 139).
All observers of energy seem to agree that various energy alternatives are virtually inexhaustible (Gordon 1981, 109).
By allocation of resources to R&D, we may deny the Malthusian hypothesis and prevent the conclusion of the doomsday models (Sato and Suzawa 1983, 81).

In the contrary view, espoused by many environmental advocates, current well-being is bought at the expense of future generations. If we do allocate more resources to R&D, and are successful at stimulating further economic growth, this will, in the environmentalist view, lead only to faster depletion, hasten the inevitable crash, and make it worse when it comes (e.g., Catton 1980). Implicit in such ideas is a call for economic *undevelopment*, for return to a simpler time of lower consumption and local self-sufficiency.

Both views are held by well-meaning persons who have intelligently studied the matter and reached opposite conclusions. Both approaches, though, suffer from the same flaw: key historical factors have been left out. The optimistic approach will be addressed first on this point, the environmental view shortly.

Economists base their beliefs on the principle of infinite substitutability. The basis of this principle is that by allocating resources to R&D, alternatives can be found to energy and raw materials in short supply. So as wood, for example, has grown expensive, it has been replaced in many uses by masonry, plastics, and other materials.

One problem with the principle of infinite substitutability is that it does not apply, in any simple fashion, to investments in organizational complexity. Sociopolitical organization, as we know, is a major arena of declining marginal returns, and one for which no substitute product can be developed. Economies of scale and advances in information-processing technology do help lower organizational costs, but ultimately these too are subject to diminishing returns.

A second problem is that the principle of infinite substitutability is, despite its title, difficult to apply indefinitely. A number of perceptive scientists, philosophers, and economists have shown that the marginal costs of research and development . . . have grown so high it is questionable whether technological innovation will be able to contribute as much to the solution of future problems as it has to past ones (D. Price 1963; Rescher 1978, 1980; Rifkin with Howard 1980; Scherer 1984). Consider, for example, what will be needed to solve problems of food and pollution. Meadows and her colleagues note that to increase world food production by 34 percent from 1951 to 1966 required increases in expenditures on tractors of 63 percent, on nitrate fertilizers of 146 percent, and on pesticides of 300 percent. The next 34 percent increase in food production would require even greater capital and resource inputs (Meadows et al. 1972, 53). Pollution control shows a similar pattern. Removal of all organic wastes from a sugar-processing plant costs 100 times more than removing 30 percent. Reducing sulfur dioxide in the air of a U.S. city by 9.6 times, or of particulates by 3.1 times, raises the cost of control by 520 times (Meadows et al. 1972, 134–135).

It is not that R&D cannot potentially solve the problems of industrialism. The difficulty is that to do so will require an increasing share of GNP. The principle of infinite substitutability depends on energy and technology. With diminishing returns to investment in scientific research, how can economic growth be sustained? The answer is that to sustain growth resources will have to be allocated from other sectors of the economy into science and engineering. The result will likely be at least a temporary decline in the standard of living, as people will have comparatively less to spend on food, housing, clothing, medical care, transportation, or entertainment. The allocation of greater resources to science of course is nothing new, merely the continuation of a two-centuries-old trend (D. Price 1963). Such investment, unfortunately, can never yield a permanent solution, merely a respite from diminishing returns.

. . . There are major differences between the current and the ancient worlds that have important implications for collapse. One of these is that the world today is full. That is to say, it is filled by complex societies; these occupy every sector of the globe, except the most desolate. This is a new factor in human history. Complex societies as a whole are a recent and unusual aspect of human life. The current situation, where *all* societies are so oddly constituted, is unique. Ancient collapses occurred, and could only occur, in a power vacuum, where a complex society (or cluster of peer polities) was surrounded by less complex neighbors. There are no power vacuums left today. Every nation is linked to, and influenced by, the major powers, and most are strongly linked with one power bloc or the other. Combine this with instant global travel, and as Paul Valery noted, "*Nothing can ever happen again without the whole world's taking a hand*" (1962, 115; emphasis in original).

. . . Past collapses, as discussed, occurred among two kinds of international political situations: isolated, dominant states, and clusters of peer polities. The isolated, dominant state went out with the advent of global travel and communication, and what remains now are competitive peer polities. . . . An upward spiral of competitive investment develops, as each polity continually seeks to outmaneuver its peer(s). None can dare withdraw from this spiral, without unrealistic diplomatic guarantees, for such would be only an invitation to domination by another. In this sense, although industrial society (especially the United States) is sometimes likened in popular thought to ancient Rome, a closer analogy would be with the Mycenaeans or the Maya.

Peer polity systems tend to evolve toward greater complexity in a lockstep fashion as, driven by competition, each partner imitates new organizational, technological, and military features developed by its competitor(s). The marginal return on such developments declines, as each new military breakthrough is met by some countermeasure, and so brings no increased advantage or security on a lasting basis. A society trapped in a competitive peer polity system must invest more and more for no increased return, and is thereby economically weakened. And yet the option of withdrawal or collapse does not exist. So it is that collapse (from declining marginal returns) is not in the *immediate* future for any contemporary nation. This is not, however, due so much to anything we have accomplished as it is to the competitive spiral in which we have allowed ourselves to become trapped.

Here is the reason why proposals for economic undevelopment, for living in balance on a small planet, will not work. Given the close link between economic and military power, unilateral economic deceleration would be equivalent to, and as foolhardy as, unilateral disarmament. We simply do not have the option to return to a lower economic level, at least not a rational option. Peer polity competition drives increased complexity and resource consumption regardless of costs, human or ecological. . . .

Peer polities then tend to undergo long periods of upwardly-spiraling competitive costs, and downward marginal returns. This is terminated finally by domination of one and acquisition of a new energy subsidy (as in Republican Rome and Warring States China), or by *mutual* collapse (as among the Mycenaeans and the Maya). Collapse, if and when it comes again, will this time be global. No longer can any individual nation collapse. World civilization will disintegrate as a whole. Competitors who evolve as peers collapse in like manner.

In ancient societies the solution to declining marginal returns was to capture a new energy subsidy. In economic systems activated largely by agriculture, livestock, and human labor (and ultimately by solar energy), this was accomplished by territorial expansion. Ancient Rome and the Ch'in of Warring States China adopted this course, as have countless other empire-builders. In an economy that today is activated by stored energy reserves, and especially in a world that is full, this course is not feasible (nor was it ever permanently successful). The capital and technology available must be directed instead toward some new and more abundant source of energy. Technological innovation and increasing productivity can forestall declining marginal returns only so long. A new energy subsidy will at some point be essential. . . .

In a sense the lack of a power vacuum, and the resulting competitive spiral, have given the world a respite from what otherwise might have been an earlier confrontation with collapse. Here indeed is a paradox: a disastrous condition that all decry may force us to tolerate a situation of declining marginal returns long enough to achieve a temporary solution to it. This reprieve must be used rationally to seek for and develop the new energy source(s) that will be necessary to maintain economic well-being. This research and development must be an item of the highest priority, even if, as predicted, this requires reallocation of resources from other economic sectors. Adequate funding of this effort should be included in the budget of every industrialized nation (and the results shared by all). I will not enter the political foray by suggesting whether this be funded privately or publicly, only that funded it must be.

There are then notes of optimism and pessimism in the current situation. We are in a curious position where competitive interactions force a level of investment, and a declining marginal return, that might ultimately lead to collapse except that the competitor who collapses first will simply be dominated or absorbed by the survivor. A respite from the threat of collapse might be granted thereby, although we may find that we will not like to bear its costs. If collapse is not in the immediate future, that is not to say that the industrial standard of living is also reprieved. As marginal returns decline (a process ongoing even now), up to the point where a new energy subsidy is in place, the standard of living that industrial societies have enjoyed will not grow so rapidly, and for some groups and nations may remain static or decline. The political conflicts that this will cause, coupled with the increasingly easy availability of nuclear weapons, will create a dangerous world situation in the foreseeable future.

To a degree there is nothing new or radical in these remarks. Many others have voiced similar observations on the current scene, in greater detail and with greater eloquence. What has been accomplished here is to place contemporary societies in a historical perspective, and to apply a global principle that links the past to the present and the future. However much we like to think of ourselves as something special in world history, in fact industrial societies are subject to the same principles that caused earlier societies to collapse. If civilization collapses again, it will be from failure to take advantage of the current reprieve, a reprieve paradoxically both detrimental and essential to our anticipated future.

References

Barnett, Harold J. and Chandler Morse. 1963. *Scarcity and Growth: The Economics of Natural Resource Availability*. Baltimore: Johns Hopkins Press.

Catton, Williams R., Jr. 1980. *Overshoot: The Ecological Basis of Revolutionary Change*. Urbana: University of Illinois Press.

Charnov, Eric L. 1976. "Optimal foraging, the marginal value theorem." *Theoretical Population Biology* 9: 129–136.

Conrad, Geoffrey and Arthur A. Demarest. 1984. *Religion and Empire: The Dynamics of Aztec and Inca Expansionism*. Cambridge: Cambridge University Press.

Dill, Samuel. 1899. *Roman Society in the Last Century of the Western Empire* (second edition). London: Macmillan.

Gordon, Richard L. 1981. *An Economic Analysis of World Energy Problems*. Cambridge: Massachusetts Institute of Technology Press.

Harner, Michael J. 1970. "Population pressure and the social evolution of agriculturalists." *Southwestern Journal of Anthropology* 26: 67–86.

Hicks, John. 1969. *A Theory of Economic History*. Oxford: Clarendon.

Krebs, John R. 1978. "Optimal foraging: Decision rules for predators." In *Behavioral Ecology: An Evolutionary Approach*, ed. J. R. Krebs and N. B. Davies, 23–63. Oxford: Blackwell Scientific Publications.

Kroeber, Alfred L. 1957. *Style and Civilizations*. Ithaca: Cornell University Press.

Lattimore, Owen. 1940. *Inner Asian Frontiers of China*. Boston: Beacon Press.

Meadows, Donella H., Dennis L. Meadows, Jørgen Randers, and William W. Behrens III. 1972. *The Limits to Growth*. New York: Universe Books.

Pfeiffer, John E. 1977. *The Emergence of Society: A Prehistory of the Establishment*. New York: McGraw-Hill.

Price, Barbara. 1977. "Shifts of production and organization: A cluster interaction model." *Current Anthropology* 18: 209–234.

Price, Derek de Solla. 1963. *Little Science, Big Science*. New York: Columbia University Press.

Renfrew, Colin. 1982. "Polity and power: Interaction, intensification and exploitation." In *An Island Polity: The Archaeology of Exploitation on Melos*, ed. Colin Renfrew and Malcolm Wagstaff, 264–290. Cambridge: Cambridge University Press.

Rescher, Nicholas. 1978. *Scientific Progress: A Philosophical Essay on the Economics of Research in Natural Sciences*. Pittsburgh: University of Pittsburgh Press.

Rescher, Nicholas. 1980. *Unpopular Essays on Technological Progress*. Pittsburgh: University of Pittsburgh Press.

Rifkin, Jeremy with Ted Howard. 1980. *Entropy*. New York: Viking Press.

Sabloff, Jeremy A. 1986. "Interactions among classic Maya polities: A preliminary examination." In *Peer Polity Interaction and Socio-Political Change*, ed. Colin Renfrew and John F. Cherry, 109–116. Cambridge: Cambridge University Press.

Sato, Ryuzo and Gilbert S. Suzawa. 1983. *Research and Productivity: Endogenous Technical Change*. Boston: Auburn House.

Scherer, Frederic M. 1984. *Innovation and Growth: Schumpeterian Perspectives*. Cambridge: Massachusetts Institute of Technology Press.

Stuart, David E. and Rory P. Gauthier. 1981. *Prehistoric New Mexico: Background for Survey*. Santa Fe: New Mexico Historic Preservation Bureau.

Valery, Paul. 1962. *History and Politics* (translated by Denise Folliot and Jackson Mathews). New York: Bollingen.

Wittfogel, Karl. 1955. "Development aspects of hydraulic societies." In *Irrigation Civilizations: A Comparative Study*, ed. Julian H. Steward, 43–57. Washington, DC: Pan American Union.

Wittfogel, Karl. 1957. *Oriental Despotism: A Comparative Study of Total Power*. New Haven: Yale University Press.

4

Less Energy, More Equity, More Time*

Ivan Illich

Unlike many students of energy and energy policy who start with the physical and then add social concerns, philosopher and social critic Ivan Illich starts with the interplay of energy, technology, well-being, and power. He derives three key propositions that relate to the premise of this book—that is, to the idea that the expected downshift in material and energy availability will coincide with an upshift in individuals' free time, an increase in positive social relations, and improved relations with the environment. Also note the convergence with Tainter's thesis on rising social complexity:

1. *Beyond a certain threshold of energy use, inequity increases, mechanical power corrupts, and time scarcity intensifies.*
2. *"The cost of social control must rise faster than total" GDP.*
3. *With appropriate technologies, "calories are both biologically and socially healthy . . . within the narrow range" of enough and not too much.*

Symbolizing Illich's findings is walking, the great equalizer, and bicycling, the single most energy-efficient mode of transit. Notice, finally, his treatment of terms like energy crisis, slave, *and* health.

It has recently become fashionable to insist on an impending energy crisis. This euphemistic term conceals a contradiction and consecrates an illusion. It masks the contradiction implicit in the joint pursuit of equity and industrial growth. It safeguards the illusion that machine power can indefinitely take the place of manpower. To resolve this contradiction and dispel this illusion, it is urgent to clarify the reality that the language

*Illich, Ivan. 1974. *Energy and Equity*. New York: Harper & Row. Excerpted and reprinted with permission.

of crisis obscures: high quanta of energy degrade social relations just as inevitably as they destroy the physical milieu.

The advocates of an energy crisis believe in and continue to propagate a peculiar vision of man. According to this notion, man is born into perpetual dependence on slaves which he must painfully learn to master. If he does not employ prisoners, then he needs motors to do most of his work. According to this doctrine, the well-being of a society can be measured by the number of years its members have gone to school and by the number of energy slaves they have thereby learned to command. This belief is common to the conflicting economic ideologies now in vogue. It is threatened by the obvious inequity, harriedness, and impotence that appear everywhere once the voracious hordes of energy slaves outnumber people by a certain proportion. The energy crisis focuses concern on the scarcity of fodder for these slaves. I prefer to ask whether free men need them.

The energy policies adopted during the current decade will determine the range and character of social relationships a society will be able to enjoy by the year 2000. A low-energy policy allows for a wide choice of life-styles and cultures. If, on the other hand, a society opts for high energy consumption, its social relations must be dictated by technocracy and will be equally degrading whether labeled capitalist or socialist.

At this moment, most societies—especially the poor ones—are still free to set their energy policies by any of three guidelines. Well-being can be identified with high amounts of per capita energy use, with high efficiency of energy transformation, or with the least possible use of mechanical energy by the most powerful members of society. The first approach would stress tight management of scarce and destructive fuels on behalf of industry, whereas the second would emphasize the retooling of industry in the interest of thermodynamic thrift. These first two attitudes necessarily imply huge public expenditures and increased social control; both rationalize the emergence of a computerized Leviathan, and both are at present widely discussed.

The possibility of a third option is barely noticed. While people have begun to accept ecological limits on maximum per capita energy use as a condition for physical survival, they do not yet think about the use of minimum feasible power as the foundation of any of various social orders that would be both modern and desirable. Yet only a ceiling on energy use can lead to social relations that are characterized by high levels of equity. The one option that is at present neglected is the only choice within the reach of all nations. It is also the only strategy by which a political process can be used to set limits on the power of even the most motorized bureaucrat. Participatory democracy postulates low-energy technology. Only participatory democracy creates the conditions for rational technology [see Dryzek, chapter 19, this volume].

What is generally overlooked is that equity and energy can grow concurrently only to a point. Below a threshold of per capita wattage, motors improve the conditions for social progress. Above this threshold, energy grows at the expense of equity. Further energy affluence then means decreased distribution of control over that energy.

The widespread belief that clean and abundant energy is the panacea for social ills is due to a political fallacy, according to which equity and energy consumption can be indefinitely correlated, at least under some ideal political conditions. Labouring under this illusion, we tend to discount any social limit on the growth of energy consumption. But if ecologists are right to assert that nonmetabolic power pollutes, it is in fact just as inevitable that, beyond a certain threshold, mechanical power corrupts. The threshold of social disintegration by high energy quanta is independent from the threshold at which energy conversion produces physical destruction. Expressed in horsepower, it is undoubtedly lower. This is the fact which must be theoretically recognized before a political issue can be made of the per capita wattage to which a society will limit its members.

Even if nonpolluting power were feasible and abundant, the use of energy on a massive scale acts on society like a drug that is physically harmless but psychically enslaving. A community can choose between Methadone and "cold turkey"—between maintaining its addiction to alien energy and kicking it in painful cramps—but no society can have a population that is at once autonomously active and hooked on progressively larger numbers of energy slaves.

In previous discussions, I have shown that, beyond a certain level of per capita GNP, the cost of social control must rise faster than total output and become the major institutional activity within an economy. Therapy administered by educators, psychiatrists, and social workers must converge with the designs of planners, managers, and salesmen, and complement the services of security agencies, the military, and the police. I now want to indicate one reason why increased affluence requires increased control over people. I argue that beyond a certain median per capita energy level, the political system and cultural context of any society must decay. Once the critical quantum of per capita energy is surpassed, education for the abstract goals of a bureaucracy must supplant the legal guarantees of personal and concrete initiative. This quantum is the limit of social order.

I will argue here that technocracy must prevail as soon as the ratio of mechanical power to metabolic energy oversteps a definite, identifiable threshold. The order of magnitude within which this threshold lies is largely independent of the level of technology applied, yet its very existence has slipped into the blind-spot of social imagination in both rich and medium-rich countries. Both the United States and Mexico have passed the critical divide. In both countries, further energy inputs increase inequality, inefficiency, and personal impotence. Although one country has a per capita income of $500 and the other, one of nearly $5,000, huge vested interest in an industrial infrastructure prods both of them to further escalate the use of energy. As a result, both North American and Mexican ideologues put the label of "energy crisis" on their frustration, and both countries are blinded to the fact that the threat of social breakdown is due neither to a shortage of fuel nor to the wasteful, polluting, and irrational use of available wattage, but to the attempt of industries to gorge society with energy quanta that inevitably degrade, deprive, and frustrate most people.

A people can be just as dangerously overpowered by the wattage of its tools as by the caloric content of its foods, but it is much harder to confess to a national overindulgence in wattage than to a sickening diet. The per capita wattage that is critical for social well-being lies within an order of magnitude which is far above the horsepower known to four-fifths of humanity and far below the power commanded by any Volkswagen driver. It eludes the underconsumer and the overconsumer alike. Neither is willing to face the facts. For the primitive, the elimination of slavery and drudgery depends on the introduction of appropriate modern technology, and for the rich, the avoidance of an even more horrible degradation depends on the effective recognition of a threshold in energy consumption beyond which technical processes begin to dictate social relations. Calories are both biologically and socially healthy only as long as they stay within the narrow range that separates enough from too much.

The so-called energy crisis is, then, a politically ambiguous issue. Public interest in the quantity of power and in the distribution of controls over the use of energy can lead in two opposite directions. On the one hand, questions can be posed that would open the way to political reconstruction by unblocking the search for a post-industrial, labour-intensive, low energy and high equity economy. On the other hand, hysterical concern with machine fodder can reinforce the present escalation of capital-intensive institutional growth, and carry us past the last turnoff from a hyperindustrial Armageddon. Political reconstruction presupposes the recognition of the fact that there exist *critical per capita quanta* beyond which energy can no longer be controlled by political process. Social breakdown will be the inevitable outcome of ecological restraints on *total energy use* imposed by industrial-minded planners bent on keeping industrial production at some hypothetical maximum.

Rich countries like the United States, Japan, or France might never reach the point of choking on their own waste, but only because their societies will have already collapsed into a sociocultural energy coma. Countries like India, Burma, and, for another short while at least, China, are in the inverse position of being still muscle-powered enough to stop short of an energy stroke. They could choose, right now, to stay within those limits to which the rich will be forced back at an enormous loss in their vested interest.

The choice of a minimum-energy economy compels the poor to abandon distant expectations and the rich to recognize their vested interest as a ghastly liability. Both must reject the fatal image of man the slaveholder currently promoted by an ideologically stimulated hunger for more energy. In countries that were made affluent by industrial development, the energy crisis serves as a whip to raise the taxes which will be needed to substitute new, more sober and socially more deadly industrial processes for those that have been rendered obsolete by inefficient overexpansion. For the leaders of people who have been disowned by the same process of industrialization, the energy crisis serves as an alibi to centralize production, pollution, and their control in a last-ditch effort to catch up with the more highly powered. By exporting their crisis and by preaching the new gospel of Puritan energy worship, the rich do even more damage to the poor than they did by selling them the products of now outdated factories. As soon as a poor country accepts the doctrine that more energy more carefully managed will always yield more goods for more people, that country is hooked into the race for enslavement to maximum industrial outputs. Inevitably the poor abandon the option for rational technology when they choose to modernize their poverty by increasing their dependence on energy. Inevitably the poor reject the possibility of liberating technology and participatory politics when, together with maximum feasible energy use, they accept maximum feasible social control.

The energy crisis cannot be overwhelmed by more energy inputs. It can only be dissolved, along with the illusion that well-being depends on the number of energy slaves a man has at his command. For this purpose, it is necessary to identify the thresholds beyond which energy corrupts, and to do so by a political process that associates the community in the search for limits. Because this kind of research runs counter to that now done by experts and for institutions, I shall continue to call it counterfoil research. It has three steps. First, the need for limits on the per capita use of energy must be theoretically recognized as a social imperative. Then, the range must be located wherein the critical magnitude might be found. Finally, each community has to identify the levels of inequity, harrying, and operant conditioning that its members are willing to accept in exchange for the satisfaction that comes of idolizing powerful devices and joining in rituals directed by the professionals who control their operation.

The need for political research on socially optimal energy quanta can be clearly and concisely illustrated by an examination of modern traffic. The United States puts 45 per cent of its total energy into vehicles: to make them, run them, and clear a right of way for them when they roll, when they fly, and when they park. Most of this energy is to move people who have been strapped into place. For the sole purpose of transporting people, 250 million Americans allocate more fuel than is used by 1,300 million Chinese and Indians for all purposes. Almost all of this fuel is burnt in a rain-dance of time-consuming acceleration. Poor countries spend less energy per person, but the percentage of total energy devoted to traffic in Mexico or in Peru is greater than in the USA, and it benefits a smaller percentage of the population. The size of this enterprise makes it both easy and significant to demonstrate the existence of socially critical energy quanta by the example of personal carriage.

In traffic, energy used over a specific period of time (power) translates into speed. In this case, the critical quantum will appear as a speed limit. Wherever this limit has been passed, the basic pattern of social degradation by high energy quanta has emerged. Once some public utility went faster than +15 mph, equity declined and the scarcity of both time and space increased. Motorized transportation monopolized traffic and blocked self-powered transit. In every Western country, passenger mileage on all types of conveyance increased by a factor of a hundred within fifty years of building the first railroad. When the ratio of their respective power outputs passed beyond a certain value, mechanical transformers of mineral fuels excluded people from the use of their metabolic energy and forced them to become captive consumers of conveyance. This effect of speed on the autonomy of people is only marginally affected by the technological characteristics of the motorized vehicles employed or by the persons or entities who hold the legal titles to airlines, buses, railroads, or cars. High speed is the critical factor which makes transportation socially destructive. A true choice among political systems and of desirable social relations is possible only where speed is restrained. Participatory democracy demands low energy technology, and free people must travel the road to productive social relations at the speed of a bicycle.[1] . . .

The typical American male devotes more than 1,600 hours a year to his car. He sits in it while it goes and while it stands idling. He parks it and searches for it. He earns the money to put down on it and to meet the monthly installments. He works to pay for petrol, tolls, insurance, taxes, and tickets. He spends four of his sixteen waking hours on the road or gathering his resources for it. And this figure does not take into account the time consumed by other activities dictated by transport: time spent in hospitals, traffic courts, and garages; time spent watching automobile commercials or attending consumer education meetings to improve the quality of the next buy. The model American puts in 1,600 hours to get 7,500 miles: less than five miles per hour. In countries deprived of a transportation industry, people manage to do the same, walking wherever they want to go, and they allocate only 3 to 8 per cent of their society's time budget to traffic instead of 28 per cent. What distinguishes the traffic in rich countries from the traffic in poor countries is not more mileage per hour of life-time for the majority, but more hours of compulsory consumption of high doses of energy, packaged and unequally distributed by the transportation industry.

Note

1. I speak about traffic for the purpose of illustrating the more general point of socially optimal energy use, and I restrict myself to the locomotion of persons, including their personal baggage and the fuel, materials, and equipment used for the vehicle and the road. I purposely abstain from the discussion of two other types of traffic: merchandise and messages. A parallel argument can be made for both, but this would require a different line of reasoning, and I leave it for another occasion.

II

Localization in Practice

If it exists, it's possible.
—Kenneth E. Boulding (1910–1993)

We in industrialized, high-consuming, growth-dependent nations are ill prepared for nongrowth, let alone energy and material descent. A common prognosis for those who foresee a resource downshift is doom and gloom. Popular authors like James Kunstler and Jared Diamond write books with apocalyptic titles (e.g., *The Long Emergency*, *Collapse*), while respected environmentalists such as James Lovelock foresee a massive culling of the global population by midcentury (i.e., in his *The Revenge of Gaia*). Such work sells, we can only guess, because the authors and their audiences cannot imagine an alternative to endless growth besides catastrophe: we either grow or we collapse, we produce more goods and services or we crawl back into the caves to shiver and perish in the dark. Perhaps the purveyors of doom and gloom believe they are serving society by frightening people into acting, by provoking them to build a future that uses less energy and fewer material resources. The psychological evidence on fear-based motivation, however, suggests just the opposite.

The key issue here is not that resources will suddenly fail us or that a long descent into the abyss is the only option. Rather what seems to be failing is imagination. We, the editors, believe that society needs to create a shared understanding of a wide range of alternative paths. The fact that human societies were once organized locally suggests that some elements of the future will be familiar. To be useful, though, localizers will have to be selective in their use of the past. The institutions and patterns of behavior that got us into the current mess (e.g., consumerism, growth-dependent economies, reliance on fossil fuels, unbridled faith in technological solutions) are certainly familiar but they are unlikely to get us out. As Boulding's quote above suggests, though, some of what once worked might work again.

The readings in this section suggest that localization does involve local provisioning and the reduction of overall community resource consumption. But these steps are not sufficient. The ultimate goal of localization is not a retreat to a narrowly defined self-sufficiency. The contemporary condition is so globally entangled that a single community's decision to voluntarily reduce its own consumption and complexity may have little positive long-term effect. As Joseph A. Tainter explained (chapter 3), a societal collapse scenario may nonetheless play out because other communities will tend to grow into the resource space created by the voluntary downshifters. Thus, even an effective simplicity movement, whether at the scale of a village, state, or country, is not a sufficient response to the situation discussed in this book. *Localization cannot be only about a specific locality.* While localization is a place-based transition, it also must include a nested set of goals, agreements, and practices supporting a regional, national, and global transition.

Proponents of positive localization also have to wrestle with prevailing assumptions about human behavior—for example, that people are predominantly greedy, hypercompetitive, individualistic, and care only about the short term. Research findings are not this myopic; it turns out that humans are perfectly capable of cooperating and sharing, self-governing, planning for the future, and exercising restraint as they respond to challenges. What's more, they innately explore new ideas and willingly share their discoveries. For several decades, social analysts, relying on theoretical models without empirical grounding, mistakenly believed all efforts at common resource management invariably ended in tragedy (e.g., Garrett Hardin's so-called tragedy of the commons). Political scientist Elinor Ostrom and her colleagues did the empirical work and developed a framework for the conditions under which humans, individually and collectively, manage resources well. That work earned Ostrom a Nobel Prize in economics.[1]

What might surprise some people is how ordinary these conditions are.[2] Further, there is abundant evidence that people across cultures and time have figured out, on their own, how to create such supportive conditions. Some of these conditions are discussed later in this book under the topic of governance (part V), but the point here is that Ostrom and her colleagues did not need to invent the principles of successful self-governance. Their contribution was to carefully analyze case studies of *existing practice* and then extract from them salient principles.

Scholars of localization might do the same. The readings presented here start the process. They provide examples of actual on-the-ground practice as well as discussions of scenarios that might emerge, soon, in localities everywhere. They allow readers to prefamiliarize themselves with promising alternatives without having to directly experience them, a process of behavior change well established in psychological research. This type of mental exploration also provides for choice: individuals can select plausible futures from among the multitude of possible scenarios, futures that make sense and do not depend on endless expansion. This process starts with envisioning but quickly proceeds to planning—that is, deciding, for instance, where and how one wants to live. With a desired and workable pattern of living in mind, one can begin to prefigure the institutions, economies, structures, norms, and behaviors that will be necessary (see Litfin, chapter 10). Finally, no one need wait to be told to start: humans have a way of starting things up without first asking permission. As energy and material constraints tighten, start-ups of all kinds, we are confident, will emerge. These readings will hopefully provide some guidance.

Notes

1. Elinor Ostrom, "The Rudiments of a Theory of the Origins, Survival, and Performance of Common-Property Institutions," in Daniel W. Bromley, ed., *Making the Commons Work: Theory, Practice, and Policy,* 293–318 (San Francisco: ICS Press, 1992); Elinor Ostrom, *Governing the Commons: The Evolution of Institutions for Collective Action* (New York: Cambridge University Press, 1990).

2. Raymond De Young, "Tragedy of the Commons," in David E. Alexander and Rhodes W. Fairbridge, eds., *Encyclopedia of Environmental Science,* 601–602 (Hingham, MA: Kluwer Academic Publishers, 1999).

5

An Arc of Scenarios*

Rob Hopkins

In this chapter, Rob Hopkins, a permaculture designer and the author of The Transition Handbook *(2008, Chelsea Green), presents a range of scenarios depicting how society might respond to a decline in net energy. Note that not all scenarios can occur because some are diametric opposites: some portray a downright dismal descent, while others look to positive futures.*

Hopkins's contribution comes in part from collecting and reviewing the work of people who are envisioning a resource-constrained future. More significantly, though, his contribution is to categorize scenarios in a coherent way (see figure 5.1) suggesting, among other things, that localization has positive features (see the box titled "Planful Shrinkage").

A key feature of scenario-based planning is that it starts with a coherent narrative about future circumstances, explores the desirability and plausibility of each scenario, and only then considers issues of design and development. Thus, analysts and practitioners start with envisioning before moving on to implementation, a sequence strongly advocated by Meadows.[1] Positive localization is an example of this combined envisioning and implementation: it visualizes a place-based response to a descent in energy and material resources and suggests a multilevel process of adaptation and innovation.

In this book, we argue that localization is a more desirable path in a time of resource descent than would be the continuation of current

*Hopkins, Rob. 2006. "Energy Descent Pathways: Evaluating Potential Responses to Peak Oil," MSc. dissertation, Plymouth, UK: University of Plymouth. Available from transitionculture.org/wp-content/uploads/msc-dissertation-publishable-copy.pdf. Excerpted and reprinted with permission.

Planful Shrinkage

The premises of localization lead to this prediction: at some time, one way or another, communities will have to downshift, like it or not, ready or not. In some places in North America, that time is already here.

Flint, Michigan, has been in decline since its dominant employer, General Motors, began withdrawing decades ago. Since then, in a desperate effort to get back to normal, Flint has tried luring GM back and attracting new industry, but to no avail. The population has dropped by nearly half, with a third put in poverty. Now civic leaders are saying it's time to prepare for a new normal. And it needn't be all bad.

Dan Kildee, a lifelong Flint resident and now Genesee County treasurer, is leading a movement to shrink Flint. This means not just operating with fewer city resources, but deliberately and systematically shrinking the city by concentrating housing, retail, and education in some areas, leaving other parts for growing food, for recreation, and for restoring natural areas.

And it's the natural areas that are really growing, enabled by the passage of a new law in Michigan. The state changed its foreclosure laws to allow county land banks to rapidly take over abandoned property. The intention was to reduce blight, a condition Flint and its big neighbor to the south, Detroit, know all too well. This permitted Flint, Detroit, and many other declining cities around the country to experiment—what we, in the book, call adaptive muddling (chapter 22). "If it's going to look abandoned, let it be clean and green," says Kildee. "Create the new Flint forest—something people will choose to live near, rather than something that symbolizes failure."

Of course, the very idea of intentional shrinkage and the relocation of people and their homes creates conflict. "Not everyone's going to win," says Kildee, "But now, everyone's losing." Localizing—whether positive or negative—is not easy but is, in many cases, unavoidable.

Source: David Streitfeld, "An Effort to Save Flint, Mich., by Shrinking It," *New York Times*, April 22, 2009.

societal trends (e.g., ever-increasing centralized power, abstract communication, unsustainable resource consumption). And, we suggest, localization is a way of managing the coming transition that may be easy to follow because, in many ways, it is familiar.

Future Scenarios

While some authors have used fiction to explore how the post-peak world might unfold (Callenbach 1975; Slonczewski 1987; Starhawk 1994; Poyourow 2005), it is from scenario planning that the most practical models have emerged. This offers a useful tool for assessing general post peak oil trends. Gallopin (2002, 365) writes "unlike projections and forecasts, which tend to be more quantitative and more limited in their assumptions, scenarios are logical narratives dealing with possibly far-reaching changes." He continues, "the scenario approach can provide a common framework for diverse stakeholders to map and address critical concerns and identify alternatives as a forum for discussion and debate" (2002). None of the authors below argue that their scenarios will unfold as described; they are storylines rather than predictions, offering tools to encourage creative thinking.

Holmgren (2005) identifies four possible scenarios. The first two, the "techno-explosion" (holidays on the moon, unlimited nuclear cold fusion etc.) and "Atlantis" (a sudden and catastrophic societal collapse), he sees as unlikely or eminently undesirable. More realistic are the third and fourth, "green-tech stability" and "Earth Stewardship." "Green-tech stability" outlines the idea that business-as-usual can continue indefinitely, with renewable energy replacing conventional energy, hydrogen cars replacing existing cars. "Earth Stewardship" Holmgren defines thus, "human society *creatively* descends the energy demand slope essentially as a 'mirror image' of the creative energy ascent that occurred between the onset of the industrial revolution and the present day" (Holmgren 2005, 7). It is this fourth scenario which Holmgren believes to "represent the only truly sustainable future" (2005). . . .

The first of Heinberg's (2004) four scenarios, "Last One Standing," describes a scenario where military force is used to secure remaining world hydrocarbon reserves. In his second, "Waiting for the Magic Elixir," a new energy source as abundant and versatile as oil is developed, such as cold fusion or the mythological "free energy." The third scenario, "Powerdown," is seen as "the path of cooperation, conservation and sharing" (Heinberg 2004, 14), a Government-led strategy utilising all the resources at its disposal to reduce per-capita consumption and build the post–fossil fuel economy and infrastructure. Finally, "Building Lifeboats," which, Heinberg (2004, 15) writes, "begins with the assumption that industrial civilisation cannot be salvaged in anything like its present form" and is a process of building community solidarity, creating a localised infrastructure and preserving and enhancing the essentials of life. Heinberg suggests that "the most fruitful response is likely to be a combination of Powerdown (in its most vigorous form) and Lifeboat Building" (2004).

The Foundation for the Economics of Sustainability (FEASTA) in Dublin developed four scenarios for Ireland's energy future. The first,

"Business as Usual," puts oil peak at 2030, with the Government doing nothing to pre-empt its arrival. The second, "Enlightened Transition," assumes that the Government decides "to use energy which is much cheaper now than it will ever be again to develop Irish energy sources and to reduce the amount of energy required to maintain and run the Irish economy" (FEASTA 2006). This results in an economy much more prepared for the peak when it does eventually arrive.

The third, "Enforced Localisation," assumes oil peak in 2007 leading to a drastic economic downturn. The economy contracts and then collapses, resulting in a very localised future, which over time becomes increasingly sophisticated, but only within much reduced energy limitations. The final one is "Fair Shares," which assumes peak oil in 2007, but a rapid Government response including the introduction of carbon rationing alongside a concerted effort to reduce energy use in all areas, and the relocalisation of most aspects of daily life. This descent is far gentler than in the "Enforced Localisation" scenario. . . .

Gallopin (2002, 383) presents three pairs of scenarios:

1. Conventional Worlds—basically business as usual, the scenarios don't deviate sharply from the present.
2. Barbarisation Worlds—like Holmgren's "Atlantis" scenario, these model a deterioration in civilisation as problems overwhelm the coping capacity of both markets and policies.
3. Great Transition—these scenarios "incorporate visionary solutions to the sustainability challenge, including fundamental changes in the prevailing values as well as novel socio-economic arrangements."

Gallopin sees the Great Transition scenario as having two possible outcomes. The first he calls "ecocommunalism," a relocalised future, with local autonomy and a dominant value of voluntary simplicity. Population contracts and cities devolve into towns and villages. The second variation on the Great Transition scenario is the "New Sustainability Paradigm." This is an almost Utopian vision wherein the global economy continues, but the culture underpinning it changes to prioritise the local, in terms of energy, production and governance. This echoes Heinberg's "Powerdown" scenario, a collective, ambitious and inclusive drive to live within the Earth's limits.

The final relevant scenarios come from outside the peak oil literature. Johnston (1991) explores paradigms for the future and identifies three possibilities. Canty (2005) observes that while Johnston does not directly address the ecological crisis, his categories are helpful. The first, "onward and upward," if translated to the environmental scenarios, is that of believing Governments and scientists will "come up with something." This echoes Holmgren's "techno-fantasy" and Heinberg's "Waiting for the Magic Elixir." Johnston's (1991) second scenario he terms the "polar view," which essentially believes that our problems will lead to "Armageddon": nuclear, environmental or economic collapse, akin to Holmgren's 'Atlantis.'

His final paradigm he calls the Evolutionary scenario, which "includes elements of each while stepping beyond either." This echoes Holmgren's "Earth Stewardship" scenario, and partly Heinberg's "Lifeboats" and "Powerdown" scenarios. He writes that most of the environmental movement operates from the polar view, arguing that we have to wake up collectively to the scale of the crisis before it is too late. This leads to solutions that promote conservation, environmental law and so on. The Evolutionary paradigm requires our evolution as a species, rather than just allowing technological solutions to "fix" the problem.

The Post Peak Scenarios Model

[Figure 5.1] was created to place the scenarios discussed [above] along a spectrum. Its form became circular rather than linear, as both extremes led to what Holmgren terms "Atlantis." Johnston's paradigms underpin the model, offering a useful template. Those scenarios between Green Tech Stability and Localisation fall within the Evolutionary Scenario, suggesting that for these to be successful, they will require our collective cognitive and behavioural evolution or adaptation as a species.

One might argue that the scenarios in the "Onward and Upward" realm require what scenario planner Pierre Wack called the "Three Miracles," namely a technological miracle (i.e., extraordinary new exploration and production levels or free/hydrogen energy), a socio-political miracle (that Government policies and cultural values will allow social exclusion to be eradicated) and thirdly a fiscal miracle, namely that the public sector will fund the implementation of that scenario (Kleiner 1996). It is the unlikeliness of all three miracles occurring that leads this dissertation to believe that the Evolutionary scenario is perhaps the most likely. . . .

Relocalisation as a Possible Response to Peak Oil

For many writers, a radical relocalisation of the economy and every aspect of life is an inevitable outcome of Peak Oil. The future, Kunstler

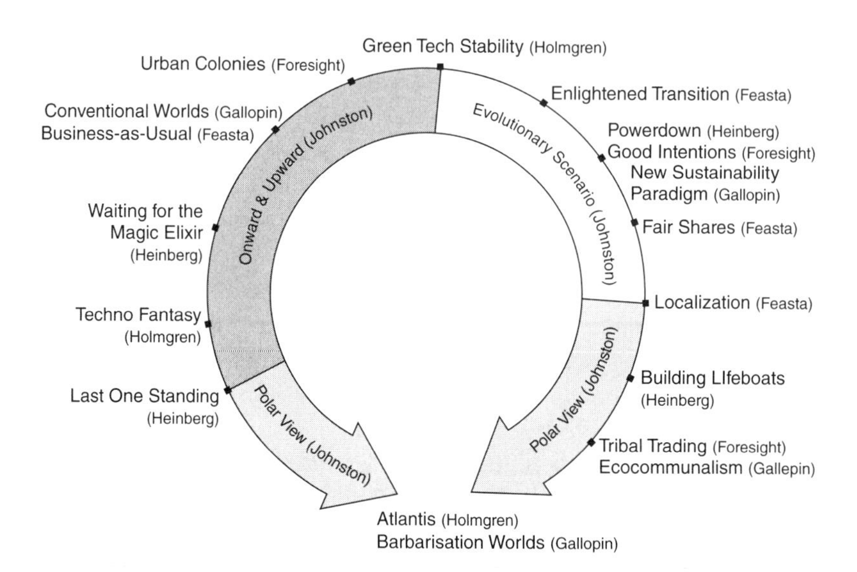

Figure 5.1
Spectrum of post-peak scenarios (Hopkins 2006; after Johnston 1991; Gallopin 2002; Holmgren 2003; Heinberg 2004; Curry et al. 2005; FEASTA 2006).

(2005, 239) believes, will be "increasingly and intensely local and smaller in scale." Indeed, Fleming (2006a, 109) states that "localisation stands, at best, at the limits of practical possibility, but it has the decisive argument in its favour that there will be no alternative." Localisation is not a concept exclusive to peak oil theorists. As [can] be seen . . . a range of arguments for more small scale, self-reliant and localised economies have been around for some time.

Peak oil has added new momentum to these calls. Fleming (2006b, 118) elucidates this argument thus:

> The transition to localisation from the global interdependence of today will be hard to achieve successfully, although it will be enforced by the breakdown of the world's energy systems and food supply. Achieving it successfully will mean establishing local political economies with an intelligence and cultural existence well in advance of the incoherent, growth-dependent and locally-atrophied market economy of our day. Doing so unsuccessfully would mean failing to build local competence of any kind, so that neither the place nor the people survive the breakdown of the global market economy which the serial traumas of energy, food, climate and social deconstruction will bring.

Ultimately, Heinberg (2004, 140) writes, "personal survival will depend on community survival," that is, individual self-sufficiency will not be a viable response to a problem of this magnitude. The first place to start in exploring the re-prioritisation of the local is with the concept of localisation.

Localisation

Norberg-Hodge (2003, 24) defines localisation thus:

> The essence of localisation is to enable communities around the world to diversify their economies so as to provide for as many of their needs as possible from relatively close to home . . . this does not mean eliminating trade altogether, as some critics like to suggest. It is about finding a more secure and sustainable balance between trade and local production.

Woodin and Lucas (2004, 69) trace arguments for economic localisation back to Keynes, who said in 1933:

> I sympathise, therefore, with those who would minimise, rather than those who would maximise, economic entanglement between nations. Ideas, knowledge, art, hospitality, travel—these are the things that should of their nature be international. But let goods be homespun whenever it is reasonable and conveniently possible, and above all, let finance be primarily local.

Kohr's (2001, 1) *theory of size*, suggests "that there seems only one cause behind all forms of social misery: bigness." Schumacher argues in favour of a higher degree of self-reliance and self-sufficiency (Vergunst 2002), not, as is sometimes misrepresented, for keeping everything as small as possible, but rather that "for every activity there is a certain appropriate scale, and the more active and intimate the activity, the smaller the number of people that can take part, the greater is the number of such relationship arrangements that need to be established" (Schumacher 1973, 54). Hines (2000a, 4) writes "localisation is not a return to overpowering state control, merely governments' provision of a policy and economic framework which allows people, community groups and businesses to rediversify their own local economies."

Shuman (2000, 46) favours the term "community self-reliance." He writes that "it suggests personal responsibility, respect for others, and harmony with nature . . . addition of the word 'community' to self-reliance underscores that the ultimate objective is a social and caring one." The term self-reliance was previously utilised by Ekins (1989, 101), who argued that it offers a key to creating sustainability: "By producing what we consume and consuming what we produce, rather than doing either through exchange, by definition we keep the externalities, positive and negative, for ourselves. The justification for so doing is clear: we will enjoy the positive externalities, rather than giving them away, and at the same time will be responsible ourselves for the negative externalities."

None of the above argues for isolationism, or for total self-*sufficiency*. Shuman (2000, 48) writes "it's easy to dismiss the principle of self-reliance by pointing to many complex products that communities cannot manufacture on their own. The goal of a self-reliant community, however, is not to create a Robinson Crusoe economy in which no resources, people or goods enter or leave. A self-reliant community simply should seek to increase control over its own economy as far as is practicable." Hines (2006b) echoes this, arguing that everything that can be produced within a nation or region should be.

Voices in favour of localisation emerge from disparate perspectives. Critics of globalisation point out that it is environmentally destructive (Retallack 2001), has disastrous effects on indigenous farmers in the developing world (Shiva 1993, 2001), promotes cultural homogeneity (Barnet & Cavanagh 2001), contributes to climate change (Sobhani & Retallack 2001), erodes biodiversity (Shiva 1993), creates inequitable trade models (Shiva 2001) and that it promotes a colonial model of development (Goldsmith 2001). Scruton (2003) argues that in the face of radical Islamic fundamentalism it is necessary to reverse the process of globalisation, and that its replacement with smaller-scale, locally based economies would facilitate the rebuilding of national security and social cohesion.

Note

1. Meadows, Donella H. (1994). "Envisioning a sustainable world." Presented at the Third Biennial Meeting of the International Society for Ecological Economics. October 24–28, 1994, San Jose, Costa Rica.

References

Barnet, R. & Cavanagh, J. 2001. "Homogenisation of Global Culture." In: Goldsmith, E & Mander, J.(eds.) *The case against the global economy and a turn towards localisation*. London: Earthscan Publications Ltd., 169–175.

Callenbach, E. 1975. *Ecotopia*. Berkeley: Heyday Books.

Canty, J.M. 2005. Environmental Healing: Shifting from a Poverty Consciousness. EcoPsychology.org. Retrieved from http://www.ecopsychology.org/journal/ezine/archive3/Environmental_Healing.pdf (Wednesday, April 26, 2006)

Curry, A., Hodgson, T., Kelnar, R. & Wilson, A. 2005. Intelligent Future Infrastructure: The Scenarios Towards 2055. Foresight, Office of Science and Technology.

Ekins, P. 1989. "Towards a New Economics: on the theory and practice of self-reliance." In: Ekins, P. (ed.) *The Living Economy: a new economics in the making.* London: Routledge.

FEASTA. 2006. Energy Scenarios Ireland. Foundation for the Economics of Sustainability. Retrieved from http://www.energyscenariosireland.com/ on Thursday, 11th May 2006.

Fleming, D. 2006a. *The lean economy: a survivor's guide to a future that works.* Unpublished manuscript, draft 26th April 2006.

Fleming, D. 2006b. *Lean logic: the book of environmental manners.* Unpublished manuscript, draft 27th July 2006.

Gallopin, C.G. 2002. "Planning for Resilience: Scenarios, Surprises and Branch Points." In: Gunderson, L.H. & Holling, C.S. *Panarchy: understanding transformations in human and natural systems.* Washington: Island Press.

Goldsmith, Edward. 2001. *The Case Against the Global Economy—and for a turn towards localization.* London: Earthscan.

Heinberg, R. 2004. *Powerdown: options and Actions for a Post-Carbon World.* New Society Publishers.

Hines, C. 2000a. *Localisation- A Global Manifesto.* London: Earthscan Publishing Ltd.

Hines, C. 2000b. "Localisation: the Post-Seattle Alternative to Globalisation." *The Ecologist Report*, September 2000, 55–57.

Holmgren, D. 2003. *Permaculture: principles and pathways beyond sustainability.* Victoria: Holmgren Design Services.

Holmgren, D. 2005. "The End of Suburbia or the Beginning of Mainstream Permaculture?" *Permaculture Magazine* 46, 7–9.

Johnston, C.M. 1991. *NecessaryWisdom: Meeting the Challenge of a New Cultural Maturity.* Berkeley: Celestial Arts.

Kleiner, A. 1996. *The Age of Heretics.* London: Nicholas Brealey Publishers.

Kohr, L. 2001. *The Breakdown of Nations.* Foxhole: Green Books.

Kunstler, J.H. 2005. The Long Emergency: Surviving the Converging Catastrophes of the 21st Century *Atlantic Monthly Press.*

Norberg-Hodge, H. 2003. Globalisation: use it or lose it. *The Ecologist.* 33(7): 23–25.

Poyourow, J. 2005. *Legacy—a story of hope for a time of environmental crisis.* Texas: VBW Publishing.

Retallack, S. 2001. "The Environmental Cost of Economic Globalisation." In: Mander, J. & Goldsmith, E. (eds.) *The Case Against the Global Economy.* London: Earthscan Publications Ltd., 189–203.

Schumacher, E.F. 1974. *Small is Beautiful: a study of economics as if people mattered.* London: Sphere Books.

Scruton, R. 2003. *The West and the Rest—globalisation and the terrorist threat.* London: Continuum Books.

Shiva, V. 1993. *Monocultures of the Mind: Biodiversity, Biotechnology and Scientific Agriculture*. London: Zed Books.

Shiva, V. 2005. *Earth Democracy: Justice, Sustainability and Peace*. London: Zed Books.

Shuman, M. 2000. *Going Local: creating self-reliant communities in a global age*. New York: Routledge.

Slonczewski, J. 1987. *A Door Into the Ocean*. London: The Women's Press.

Sobhani, L. & Retallack, S. 2001. "Fuelling Climate Change." In: Mander, J. & Goldsmith, E. (eds.) *The Case Against the Global Economy and for a turn towards localisation*. London: Earthscan Publishing, 224–241.

Starhawk. 1994. *The Fifth Sacred Thing*. New York: Bantam Books.

Vergunst, P.J.B. 2002. The Potential and Limitations of Self-reliance and Self-sufficiency at the Local Level: views from southern Sweden. *Local Environment* 7(2): 149–161.

Woodin, M & Lucas, C. 2004. *Green Alternatives to Globalisation—A Manifesto*. London: Pluto Press.

6

Inhabiting Place*

Robert L. Thayer, Jr.

Landscape architect Robert Thayer argues that the present order—globalized and specialized—is abnormal. What sustains humans, Thayer writes, "are finite territories inhabited by small bands of humans." And, he finds, new bands are forming all over North America in what he calls bioregions. Whereas many forces of modern society feel like "something that is 'being done to us' and out of our control," localization is internal to our place and something we can indeed influence. Notice that Thayer is saying that humans have always been place based; they are so today and they will become increasingly more so as localization shifts attention toward the life place.

No real public life is possible except among people who are engaged in the project of inhabiting a place.
—Daniel Kemmis, 1990

There is no such thing as a citizen of the world.
—Manuel Castells, 1997

The year is 1999. When I arrive slightly tardily to the Regional California Fish and Game Headquarters in Yountville, Napa watershed, the meeting is standing-room only: at least twenty people are in chairs squeezed together arm to arm around the table and an equal number are seated or standing at the perimeter. Incongruously, a stuffed polar bear, moose, and caribou peer upon the assembled crowd from outside the glass-walled entrance of the room, adding irony to this most regional

*Thayer, Robert L. Jr. 2003. "Reinhabiting: Recovering a bioregional culture." In *LifePlace: Bioregional Thought and Practice.* (c) 2003 by Robert L. Thayer, Jr., 52–70. Published by the University of California Press, Berkeley, CA. Excerpted and reprinted with permission.

assemblage of voluntary participants—the newly forged Blue Ridge–Berryessa Natural Area Conservation Partnership. Our common bond this Friday, like that of a dozen or more monthly Fridays before it, is four hundred thousand acres of mountainous terrain covered by oak woodland, grassland, chaparral, serpentine outcrop, and creekside riparian lands. The "BRBNA" (Blue Ridge–Berryessa Natural Area) is the awkward moniker arrived at by committee as a label for the wild, interior Coast Range lands extending southward from the Mendocino National Forest nearly to Interstate 80 between San Francisco Bay and Sacramento. Stretched between the upper watersheds of Putah and Cache Creeks, BRBNA is an extraordinary mosaic of spectacular cattle ranches, nationally significant wildflower displays, rugged whitewater creek canyons, two water supply reservoirs (one ringed with double-wide trailers and dotted with jet-skis and fishing boats), an enormous open-pit gold mine, a university wildlands research center, a secluded, clothing-optional hot springs resort, a federally designated wilderness study area home to bald eagle and tule elk, and vast expanses of scrubby chamise chaparral so dense that a human could not penetrate more than a foot into its midst. In spite of its location in the California "Coast" Range (a misnomer), this is true, nearly arid, remote, wild, *western* land that Zane Gray or Louis L'Amour would have loved. There is something for everyone in the BRBNA, and nearly everyone has come to this meeting, including the politicians. The meeting agenda is full and proceeds with a report of the Bureau of Reclamation's concession services plan for the resorts around Lake Berryessa; news of the acquisition of the twelve-thousand-acre Payne Ranch by the Bureau of Land Management; talk of the possibility of a six-million dollar congressional budget line item request to the House Interior Subcommittee for conservation land acquisitions; and vocal concerns over threats to biodiversity from a proposed "Moonie" (Unification Church) community on the BRBNA periphery. The agenda closes with a progress report on the transfer of Homestake Mine land to the university's McLaughlin Ecological Reserve. As items are discussed, the eyes of the local politicians widen; it seems they have rarely seen private ranchers, environmentalists, agency folk, professors, and game wardens around a table free of dispute or conflict. We leave the meeting with a heady, optimistic feeling.

The BRBNA, which is the brainchild of Ray Krauss, Homestake's former environmental manager, is one of several "bioregional" groups that have emerged in the Putah and Cache watersheds. Yet the . . . volunteer domains of concern here show but a fragment of a growing phenomenon: all over North America people have begun to assemble on behalf of various watersheds, coastlines, mountain ranges, prairies, and lake regions. It is as if some long-lost ingredient of the national character has suddenly been rediscovered, a vital puzzle piece of democratic community in the midst of the myth of American individualism. Whatever the reason, people are now forming shared "communities" of mutual concern and action around the "tables" of natural regions. As the author and progressive politician Dan Kemmis might say, community is forming around the politics of place.[1]

What drives the human predisposition to gather in small groups that identify strongly with naturally definable regions, and to consider and participate in the best ways to guide those regions' futures? I cannot be certain, but I suggest a hypothesis: the newly globalized and highly specialized society in which we now find ourselves embedded is not the evolutionary norm; rather, what sustains us are finite natural territories inhabited by small bands of humans. We establish groups working on behalf of river basins or mountain ranges simply because it feels quite natural for us to do so. This chapter builds a bridge between those first people who originally inhabited a life-place and today's residents who seek to "reinhabit" theirs in a manner befitting the place.

Life-Place and Human Evolution

For most of our existence on earth, *Homo sapiens* banded together cooperatively to sustainably harvest the natural potentials of finite territories. Perception of the extent of the world and the size of its communities matched the ability of a particular group to derive livelihood from its world. Thus, the evolutionary survival of humanity has depended largely upon *social cooperation in place*. As late as 1981, economist Hazel Henderson pointed out that most of the world's people were still sustained by "growing their own food, tending their own animals in rural areas, and living in small, cooperatively run villages and settlements or as nomads following herds, harvesting wild crops, fishing, and hunting in economies based on barter, reciprocity, and redistribution of surpluses according to customs."[2] The story of humanity, of course, is not finished; and in its latest chapters it has begun to diverge from this theme. While humans have flourished as economics and technology have allowed a certain transcendence of time and place, the fate of many other characters in the story—our companion species—is less certain.

Raymond Dasmann differentiates between *ecosystem people* and *biosphere people*, the former being those who live within the ecological limitations of their home area in order to survive and the latter being those tied into the global economy, whose livelihood is not necessarily dependent on the resources of any one particular region.[3] As tool use, long-distance trading, communication, and technical dependency evolved, human existence turned away from regional ecosystems toward the modern "biosphere-based" condition. In the process many regions became highly dependent upon imports from remote places, and to pay for those imports, particular commodities were harvested far in excess of regional carrying capacities and exported out of home regions in exchange for needed currencies. This extended the geographic range of human impact well beyond the limits of immediate human perception. In recent history, the main preventatives for the collapse of modern biosphere-based cultures have been the widespread utilization of military force, the accelerated creation of new technologies, and the exploitation of nonrenewable fossil fuels. To those three characteristics of modernity, technophiles now add a fourth: information, which some advocates suggest is an equivalent partner to matter and energy among essential qualities of the universe. Information and the technologies that store and transmit it seem to have become emblematic of our modern biosphere-based society. Yet ecosystem people dealt with matter, energy, and information in ways that were equally powerful and that allowed them to thrive in their regions for thousands of years.

Place, Language, and Culture

Language is one good indicator of life-place boundaries. By the time of European settlement, the Putah-Cache Creek region and its human inhabitants were a small subsection of what some anthropologists have called the "central California cultural climax"—a presumed territory and quality of existence where human habitation was in sufficient balance with its surroundings to achieve a significant population density with little apparent detriment to the carrying capacity of the enveloping region.[4] The Sacramento Valley contained a culture with a common linguistic heritage (Penutian) and widely shared ritual practice yet marked by very localized dialects. These dialects are a consequence of California's unique geography, in which the "warp" of its main ecosystems is transected by the "weft" of its streams and rivers. The major ecoregions of

California are primarily determined by latitude, elevation above sea level, landform, and distance from the ocean and its prevailing winds and currents. California bioregions are more lush at higher elevations, higher latitudes, and locations closer to the ocean, while lower elevations, lower latitudes, and areas remote from maritime influences are more arid. Vegetation zonation in California stratifies according to altitude bands, moving from low-elevation desert and grassland in the interior to semiarid oak savanna foothills, then to mixed forests at middle elevations, and then to more moist coniferous forests at high elevations. As they descend rapidly from mountains to the ocean, California streams and rivers cut across these ecosystems. Watersheds in California, therefore, are not necessarily coincident with bioregions.[5] . . .

The River Patwin tribes who lived adjacent to the Sacramento River and its seasonal floodwaters had access to vast schools of salmon, steelhead, and other anadromous fish, and these harvests provided a considerable portion of their diet. People on tributaries such as Putah and Cache Creeks, which prior to upstream impoundment and twentieth-century flood control structures merged with the Sacramento floodplain marshes mainly during high water, found at least enough salmon and sturgeon to provide auxiliary food. For these groups, a favorite way to prepare and preserve salmon was to dry the meat and pulverize it into a flourlike powder, which then could be easily carried, stored, or mixed with other foods.

For the Sacramento Valley bioregion as a whole, including the lands extending up low foothill tributary streams both east and west, the major food source by far, and the single most important bio-indicator of culture, was the acorn. Starting about two thousand years ago, the flat stone *metates* and *manos* of the inhabitants of this bioregion, used to process grass seed, were gradually and almost entirely replaced by the round, stubby pestles and bedrock mortars more appropriate for releasing the many calories available in the oilier acorns. Acorns figure significantly in the geography of native Californians; there is a modest congruity among oak distribution, regional boundaries of the Penutian languages, the drainage basins of the Sacramento and San Joaquin Rivers, and the territorial extent of the central California "climax" cultures characterized by semisubterranean dance lodges, spirit impersonation, and secret societies. These four extents do not precisely coincide, but there is sufficient overlap to suggest a general life-place relationship among landform, watershed, dominant foods, major linguistic patterns, and spiritual and cultural practices.[6]

The Cultural Hypothesis

As distinctly unique associations of plants and animals defined general biological regions and as these regions were further dissected by watershed corridors, original human cultures adapted to these patterns in close relation to natural boundaries. Although the cultural ecology of acorn-eating peoples varies considerably, the California acorn-dependent cultures are far less distinct from one another than they are from the northwestern salmon-based cultures, the southwestern cultivators, or the Great Basin hunter-gatherers.

The spatial relation of indigenous peoples to the regions they occupied suggests a possible "cultural hypothesis" about bioregions. In contemporary terms, bioregions, or life-places, are an alternative geography for humans that recognizes the limitations and potentials of the immediate regions in which people live and localizes the affections and actions of inhabitants in a manner that is socially inclusive, ecologically regenerative, and spiritually fulfilling. In short, an overall Cultural Life-Place Hypothesis might be summed up as follows: *Human culture is best suited to naturally defined regions and reasonably sized communities. Bioregions, or life-places, are the evolutionary norm, not the exception.*

I now live within the memory-space of a formerly bioregional culture. I reflect upon these first peoples with an eye to understanding their response to our region and hold forth the hope of emulating their lessons in this bioregion once again. With unsentimental reason and respect, might we learn from first peoples how to share a mutual community of reciprocity between human and nonhuman life?

Globalism and Its Discontents

All higher species must perceptually distinguish figure from ground and determine wholes from collections of parts. Humans are perhaps the most highly skilled animals at pattern recognition, and the life-place concept is a natural extension of our proclivity to sense patterns—to uncover, in holistic fashion, the necessary "units" of the living environment by which humans and other species are able to survive. For example, early humans needed not only to discriminate between similar-looking edible and toxic plant species but also to assemble the various "parts" of the environment into constructs that afforded them opportunities to find edible plants in the first place. This required an ability to associate individual species with combinations of major landforms and other physiographic characteristics.[7]

However, this ability to discriminate has been taken to an extreme in the last century or so. In the course of evolution, humans have relied on both "lumping" and "splitting" skills to survive, but dissociation of parts from the whole, minute examination of certain parts, and the reassembly of parts via technology without consideration of proper context are today the norm. Much of supposedly "objective" reality is considered mere social construction. While there is a kernel of truth to the notion that "nature" itself is just such a social construct, it is possible to take this line of reasoning too far. Without a tangible, *grounded* (in the literal sense) basis, we lose all connection to nature or bioregion "constructions" that provide us with our own, necessary, and proper context. The recognition of a life-place, or bioregion, then, is perhaps an acceptance of the need for us all to reassemble the world by integrating the natural dimensions of each of its various regions with a deepening sense that we *inhabit* a specific place.[8]

Ironically (or perhaps perversely), as scientists continue to "split" the world of knowledge into ever more narrow specializations, accelerating globalization would have us believe that the world is shrinking and becoming one homogeneous culture. Is it? The ubiquity of instantaneous telecommunication, the emergence of English as a world language, the dominance of American pop culture, and the unchecked explosion of capitalism and corporatism might make it seem as if the human proclivity for lumping had at last won out. This interpretation, however, glosses over the deep differences in the nature and culture of humanity. Reassembling a world dissociated by industrial technology and scientific reductionism by means of electronic/capitalistic hegemony strikes many as a cure worse than the disease itself.

In his sequential books *The Network Society* and *The Power of Identity*, sociologist Manuel Castells diagnoses the phenomenon of technologically enabled cultural globalism.[9] As Castells observes, in an emerging global network society characterized by virtual reality, rapid information, blurred social spaces, dissolution of the idea of time, accumulation of wealth by the few, and social arrhythmia in the familiar cycles of human life, power is being reorganized from the "space of places" to the "space of flows."[10] But he also notes the emergence of many powerful communal resistance identities, each rallying around a particular value, such as religion, state, region, neighborhood, tribe, family, sexual orientation, or environment. These resistance identities do not fit logically together, nor do they act in consort; in fact, many are totally unrelated to one another or even diametrically opposed. Resistance identities are, however, all communal: they define exclusive communities of resistance to the perception or action of external oppression from the dominant social structure—a process that Castells describes as "the exclusion of the excluders by the excluded."[11] A resistance identity may draw cultural "boundaries" around itself, and within that defined "territory" (whether ideological, geographical, or both) it inverts the guiding premises and expected behaviors presumed by the major social paradigm.

For example, the widespread hegemony of technology presupposes that everyone should acquire computer hardware, software, and requisite skills as soon as these become available. Countering this presumption is a neo-Luddite resistance movement that legitimizes for its participants the act of not buying, not using, or even destroying computers. While the objects of various resistance identities may be different, the processes of carving out ideological territory are similar, whether the resistance identity in question is Islamic fundamentalism, the militia movement, the animal rights movement, the environmental movement, or the feminist movement.

An irony of the current state of civil society emphasized by Castells is that individuals who seek to establish their identity and engage in social action are not likely to do so as members of a "global network society." Rather, they are more likely to participate in various organizations in resistance to it. Notwithstanding the fictitious actors in slick TV ads for dot-coms or the sullen hordes of laptop-clicking, cell phone–calling business folk working their way through the airport hubs of the transnational corporate world, few people overtly identify themselves as champions of the "global network society" or wear its emblems. It is as if the network society were a construction built of corporate advertising hype featuring syrupy images of folks chatting happily via Internet and cell phone across continents and cultures. Perhaps the most prevalent feeling about globalism, whether one agrees with it or not, is that it feels external, something that is "being done to us," or at least something that is proceeding without our input or control.

So, as Castells elaborates, it is by means of the resistance identities to network society that people most frequently identify themselves in the arena of social action:

> Thus, social movements emerging from communal resistance to globalization, capitalist restructuring, organizational networking, uncontrolled informationalism, and patriarchalism—that is, for the time being, ecologists, feminists, religious fundamentalists, nationalists, and localists—are the potential subjects of the Information Age.[12]

I also note, with great interest, the word *localists* in Castells's argument. For the first time in history, a world of subjects (i.e., social actors constructing an identity) is being shaped in part by spatial decentralization while the dominant technical paradigm races toward consumerist homogeneity and corporate economic concentration. If globalism is so widespread and so inevitable, why don't more people overtly embrace it? Society, it seems, has never been in this "place" before.

Castells also suggests that national governments have become increasingly obsolete as mediators between the global culture-economy and the more localized, specialized communities of resistance; territorial identity and the worldwide resurgence of local and regional movements indeed foreshadow the "reinvention of the city-state as a salient characteristic of the new age of globalism."[13] Unlike past decentralization following the collapse of empires, this form of decentralized social identity is driven by and is in direct opposition to economic centralization. In this strange postmodern condition, the idea of literal and figurative common ground on which a culture can aggregate is the subject of considerable debate. We now enter truly uncharted cultural territory, where a strange admixture of global and local identities pulls us to and fro. As the globe "shrinks" and becomes more "accessible," so, too, does our social resistance increase and our affinity for the local deepen. In both culture and geography, there is something inherent in humanity that does not want us to become one.

Reinventing Common Ground

. . . In a world of broad corporatist networks of special interest, Saul reasserts the need for specific communities of dis-interest. In Saul's conception, far from being apathetic, a community of dis-interest is a collective manifestation of civic duty practiced in a specific place *without* expectation of personal gain—like serving on a jury. Jurors have no stake in the outcome of their deliberations, but they participate out of a sense of contribution to the local practice of democracy.[14] . . .

. . . Daniel Kemmis . . . in his seminal book *Community and the Politics of Place*, notes how people occupying the same geographical region seem trapped by their so-called public posturing to endorse either the myth of rugged individualism or the mire of regulatory bureaucracy in choosing sides during land use conflicts. Meanwhile, the shared values of place and region are ignored. Kemmis advocates a return to republican (small r) values. His view is that government should facilitate the *best* of human civic behavior rather than the worst—bureaucratic insularity, confrontational stalemate politics, fear of litigation, or public "hearings" where no one listens.[15]

To . . . conclude . . . the *role of place and region is vital to the politics and culture of a democratic community*. As Kemmis emphasizes, civic participation needs a tangible object—a sort of "table" around which the "*res*, the public thing of the 'republic' . . . could gather us together and yet prevent us from falling over each other."[16] This tangible object is the shared *place* itself, which is to say, the community, the bioregion, the *life-place:*

> That we inhabit a global economy has become commonplace. What is not so universally understood is that the organic integration of the global economy is drawing into play suborganisms that refuse to be ordered by anything other than their internal logic.[17]

As the long-entrenched politics of left versus right, individual freedoms versus heavy-handed government, and "Wise Use" rhetoric versus environmental monkey-wrenching are caught up in the rush of globalization, a political vacuum is created. That vacuum draws into itself the possibility of a new politics focused on region, community, and identity—a place-bounded resistance identity capable of transcending bipolar politics in favor of regenerative civic democracy.

Today, however, a new equilibrium is being reached between communities of *interest*, which tend toward the global, and communities of *place*, which tend to be local. Although the bioregional movement traces its roots back to radical social theory and early left-wing environmentalism, the modern move toward equilibrium is being driven as well by the social experimentation embodied within ecosystem management, place-based civic democracy, ecologically based regional planning, alternative economic theory and practice, and a host of related "relocalization" efforts. This has resulted in something uniquely absent from the typical liberal-conservative spectrum. Grassroots, multistakeholder efforts on behalf of natural regions or watersheds have been labeled "Wise Use"/industry scams by the green left because of their alleged efforts to dupe locals and violate federal land management policies as often as they have been branded government "enviro" conspiracies by the right wing for their supposed assaults on personal freedoms and individual property rights. The truth, which neither the traditional right nor left wishes to admit, is that broadly enfranchised, local, grassroots efforts to identify with and care for natural regions are so powerful, so ultimately democratic, and so basically popular with the American people that they threaten the huge, entrenched political organizations on both sides.[18]

The Nature of Life-Place Culture

The failure of the traditional government-agency, single-resource approach to meet multiple resource management needs has led to a considerable broadening of the cultural assumptions of the "original" bioregionalists. When coupled with emerging trends in ecosystem management, regional planning, grassroots bioregionalism, alternative economics, and certain strains of social criticism, this multidimensional response suggests a cluster of descriptive and prescriptive principles that could begin to define a life-place culture. As people from more sectors of society and the economy are negatively affected by global trends, a convergence of local interest on the life-place seems inevitable. When valley farmers can no longer find buyers for their crops of apricots or tomatoes—crops that have helped define the region for decades—they may discover new friends, and, perhaps, even new markets, among their nonfarming neighbors. When energy users in California realize they are hamstrung by out-of-state energy suppliers, price manipulators, and regulators, they may turn instead to more local energy solutions. When formerly adversarial groups find themselves, for better or worse, inhabiting the same bioregion, and facing the same limits and potentials, an embryonic life-place culture may arise. How might a seasoned life-place culture be characterized? First and foremost, it would be a collective human endeavor. In addition, it would be

- framed by the nature of the region (identifying with and growing more attached to place)
- concerned with all life, human and nonhuman
- scaled to territories comprehensible to human perceptions, affections, and activities
- focused upon or catalyzed by tangible objects of shared social and natural value (watersheds, species, habitats, disenfranchised groups)
- based on face-to-face communication in real time and space enriched through horizontal networks of civic engagement built on mutual trust (neighborliness) in spite of differences of opinion
- grounded in respect for and dependence on local wisdom and knowledge
- balanced between freedom and obligation (negotiating a middle path between annihilation of open country at one extreme and eco-monkey-wrenching at the other)

- supported equally by common sense, creativity, ethics, intuition, memory, and reason
- enfranchising all potential “stakeholders” equally
- equitable and socially just, featuring symmetrical power arrangements
- capable of creating social capital, or building “capacity” for problem solving, among a broad base of citizenry
- innovative in establishing institutional cooperation and horizontal linkagcs
- reinhabitory, or invested in the future (fostering life as though one’s future grandchildren would be living in the same place and doing the same things)
- as supportive of communities of place as it is of communities of interest
- based on quality of life over time, including the means of making a living in place
- regenerative (careful to perpetuate valued social institutions, ecosystems, and physical/natural resources over the long term)
- respectful of natural boundaries and systems that often straddle illogical political demarcations
- evolutionary (capable of being “grown” over time, rather than being forced upon or superimposed over existing political frameworks)
- adaptable to change from without or within

A life-place culture, then, is an alternative mode for contemporary humanity that recognizes the limitations and potentials of the immediate regions in which people live and strives to relocalize the affections and actions of inhabitants in a manner that is socially inclusive, ecologically regenerative, economically sustainable, and spiritually fulfilling. The culture of reinhabitation is life-place culture: the rediscovery of a way to live well, with grace and permanence, in place.

Charmed by a Stone

In April 1991, three friends and I ride bicycles in the sixty-mile Tour of the Lost Valley, our mid-forties age feeling like twenty to us as we climb the narrow, winding tarmac from Williams toward Lodoga, in the west-side foothills. It is a cool, sunny day following a rainstorm, and the foothills are as intensely mint-green as could ever be imagined. While resting at the top of the big climb, wolfing down energy bars and chugging polyethylene-bottled water, we are pleasantly buzzed by a golden eagle. To the east, the patchwork valley stretches out to the Sierra foothills, and to the west lies a rolling green carpet of grasses, blue oaks, and wildflowers that could serve well as an official billboard for the "real" California.

By this year and this ride I am thoroughly caught up in a search for knowledge of my home region. I have explored the innermost Coast Range foothills by foot, bicycle, canoe, and automobile. After eighteen years, I feel completely at home amid the agricultural environment of the valley. Gradually I have come to realize that, in contrast to what I believed upon my immediate arrival here, there has been an abundant native population in this place, and *not* all indigenous California people died in the disease epidemics of 1833 and 1834. Perhaps I am typical of many other white Americans: vaguely aware that primal peoples once lived on "my land" but rather ignorant of who they were, exactly where they were, *how* they lived here, and most of all, whether any are *still* here.

On that April day I do not recognize or know the significance of a small object that I pick up from the ground while resting against my bicycle. It is made of hard, black-and-white-flecked stone—something like granite. It appears to be manmade, uniformly round in cross section yet tapered at both ends. I ponder it briefly, thinking it may be a sample intended for testing the strength of rock, or perhaps a balustrade from an ornate stone garden fence. I put the stone in my bike jersey pocket, complete the ride, and, upon arriving back home, stow the object away in a "junk" drawer.

Three years later, while researching the first peoples to live in this region, I see with astonishment in an anthropological text an illustration of the exact object that I found: a four-thousand-year-old "early horizon" steatite charm stone. The function of such stones is still debated, but charm stones were found in native graves dating up until a thousand years ago. The one I found precisely matches the form and material of charm stones found in the earliest horizon of archeological exploration. It is thought that such charm stones were suspended over spots in the stream to "charm" the fish into being caught. My stone, I suspect, accompanied a load of local gravel brought to buoy up a new asphalt "river." Looking at the stone, I imagine it hanging vertically from a branch and wonder which species of fish it was intended to catch, which hands so carefully hewed it out of a larger piece of stone, what the particular worldview of the individual who made it was, and whether, perhaps, that person wondered, as I do now, who had come before him in that place.

For those of us come only recently to a territory, it is difficult to imagine the hundreds of generations that have passed down intimate knowledge on how to live there: fathers and mothers teaching daughters and sons the best means of surviving and thriving in this place; whole communities of humans and nonhumans so entwined with the land that any slim boundaries between self and other, sky and earth, water and soil, animal and human, must have been inconceivable. How could the land not have been sacred to them? What else could explain the persistence of people in place over such a long time that entirely separate languages evolved within distinct watersheds draining only a few thousand hectares?

The charm stone now sits in a leather pouch upon a small meditation altar in my home office. It symbolizes for me an acknowledgment that we are all "dancing on sacred land." Over the years, I have made it a hobby to piece together every shred of information possible on these indigenous peoples who were here long before me—where they lived, what languages they spoke, what foods they gathered, what fish they caught, what animals they hunted, what gods and spirits they beckoned, what dances they danced—and where they are now. Can we use the echoes of their culture to help us reassemble and reinhabit our fragmented world?

Notes

1. Daniel Kemmis, *Community and the Politics of Place* (Norman: University of Oklahoma Press, 1990).

2. Hazel Henderson, *The Politics of the Solar Age* (New York: Anchor Press, 1981), 25, quoted in Raymond Dasmann, *Environmental Conservation* (New York: John Wiley, 1984), 412.

3. Dasmann, *Environmental Conservation,* 412.

4. See Alfred Kroeber, *Area and Climax,* University of California Publications in American Archeology and Ethnology, vol. 37 (Berkeley: University of California Press, 1936).

5. See James Omernik and Robert Bailey, "istinguishing between Watersheds and Ecoregions,"*Journal of the American Water Resources Association* 33 (1997): 935–949 for a discussion of the differences between ecoregions and watersheds.

6. For Penutian language distribution, see Alfred Kroeber, *Handbook of the Indians of California* (1925; reprint, New York: Dover, 1976); for acorn distribution, see J. R. Griffin and W. D. Critchfield, *The Distribution of Forest Trees in*

California, U.S. Forest Service Research Paper PSW 82 (Berkeley, Calif.: Pacific Southwest Forest and Range Experiment Station, 1972).

7. See J. J. Gibson, *The Ecological Approach to Visual Perception* (Boston: Houghton Mifflin, 1979).

8. See Neil Everenden, *The Social Construction of Nature* (Baltimore: Johns Hopkins University Press, 1992).

9. Manuel Castells, *The Rise of the Network Society,* vol. 1 of *The Information Age: Economy, Society and Culture* (Malden, Mass: Blackwell, 1996), and *The Power of Identity,* vol. 2 of *The Information Age: Economy, Society and Culture* (Malden, Mass.: Blackwell, 1997).

10. Castells, *Rise of the Network Society*, 378.

11. Castells, *Power of Identity*, 9.

12. Ibid., 360–361.

13. Ibid., 357.

14. John Ralston Saul, *The Unconscious Civilization* (Concord, Ontario: Anansi, 1995).

15. See Kemmis, *Politics of Place;* also Daniel Kemmis, *The Good City and the Good Life* (Boston: Houghton Mifflin, 1995).

16. Daniel Kemmis, "A Democracy to Match Its Landscape," in *Reclaiming the Native Home of Hope: Community, Ecology, and the American West,* ed. Robert B. Keiter (Salt Lake City: University of Utah Press, 1998), 7.

17. Ibid.

18. Mark Nechodom, "Democracy, Ecology and the Politics of Place" (Ph.D. diss., University of California, Davis, 1998); Doug Aberley, "Interpreting Bioregionalism: A Story from Many Voices," in *Bioregionalism,* ed. Michael McGinnis (London: Routledge, 1999), 13–42; Michael V. McGinnis, ed., *Bioregionalism* (London: Routledge, 1999); and Timothy Duane, *Shaping the Sierra* (Berkeley: University of California Press, 1998), all address the relationship between social theory and grassroots action in ecosystem management and bioregionalism.

7

Locally Owned Business*

Michael H. Shuman

Economist, writer, and consultant Michael H. Shuman compares two models of business and, by implication, of an economy: locally owned and import-substituting (LOIS) and there-is-no-alternative (TINA). The imports he refers to are the goods and services that come into a community. The consequence of such importing is the exiting of wealth from that community.

While it may seem that American businesses largely exist on the scale of Wal-Mart, General Motors, Citigroup, and Google, Shuman points out that "at least three-quarters of our economic activity is currently place based." Thus, he argues, local business already exists, and more is possible (see the box titled "Localizing Finance"). The problem is that local officials often believe there is no alternative to big business.

Notice that Shuman's thesis is not antitrade nor is it a strict localism. Rather, he says that communities of locally owned businesses should be self-reliant, meeting as many of their "needs as possible, then [competing] globally with a diversity of products." This is consistent with a theme of this book: localization is not strictly about the local; it is about orienting attention, commitment, production, and consumption to the local.

Finally, notice that Shuman's definition of local is primarily geographic and economic. Other authors in this book stress the ecological, social, and psychological. Combined, one gets a sense of the richness of both existing and potential localizing practices.

The Hershey Chocolate Company in Hershey, Pennsylvania, is not your typical LOIS [locally owned and import substituting] business. Its stock

*Excerpted and reprinted with permission of the publisher. From *The Small-Mart Revolution: How Local Businesses Are Beating the Global Competition*, copyright © 2006 by Michael H. Shuman, Berrett-Koehler Publishers, Inc., San Francisco, CA. www.bkconnection.com

Localizing Finance

Localizing can occur in some of the unlikeliest places. Texas, a state that virtually epitomizes the American credo of rugged individualism, free markets, economic growth, and capitalism, is seeing a "bank local" movement. "Real Texans bank locally," says the advertisement for Texas Dow Employees Credit Union, which also suggests that out-of-state banks "head back home and make their profits where they live."

The financial collapse, taxpayer bailouts, and subsequent profits and bonuses on Wall Street have opened up marketing space for community banks. Small local banks—some 8,000 nationwide—don't have the political power and capital of the big banks, but they do have the trust and accountability that savers and borrowers want. A Gallup poll in June 2009 found that only 18 percent of respondents had high confidence in the nation's big institutions. "I can't beat Wells Fargo and Bank of America nationally," says the Texan Edward Speed, who runs Texas Dow Employees Credit Union, "but I can certainly beat their branch across the street."

Perhaps most surprising, credit unions and private banks are collaborating to win back market share from centralized banking institutions. More than 300 local banks met in Dallas in August 2009 to market themselves collectively.

In those campaigns, they can credibly claim to be the most stable of banks. The Federal Deposit Insurance Corporation, the federal agency that insures deposits, found that banks with less than $1 billion in assets were the best capitalized. That is, they kept the most capital in reserve as a cushion against downtowns. During the 2008 financial crisis, the small banks kept lending while the big ones stopped (despite collecting bailouts from taxpayers).

There is probably no better symbol of the meaning of *local finance* than an advertisement seen above a Denver baseball game in April 2009. A tiny, single-engine propeller airplane sputtered across the sky towing a banner for a local bank that read "This is the closest thing we have to a private jet."

Source: Zachery Kouwe, "Small Banks Move In as Giants Falter," *New York Times*, November 2, 2009.

is publicly traded, which normally makes local ownership impossible, but a local charity, the Hershey Trust, keeps ownership local by controlling 77 percent of all voting shares.[1] Unlike most LOIS businesses, it is hardly small. In 2001 about sixty-two hundred employees were on payroll, many living in the Hershey area. The company not only saturates local chocolate demand but also sells worldwide to the tune of $4.6 billion per year.

The Hershey Trust is effectively the heart that pumps monetary blood throughout the regional economy. It owns 100 percent of the shares of the Hershey Entertainment and Resorts Company, which employs another fifteen hundred locals (plus five thousand seasonal workers) in its amusement parks, stadiums, campgrounds, country clubs, and numerous other enterprises. On top of that, the Trust runs a school for twelve hundred underprivileged kids, grades K through 12. Milton Hershey put all his stock in the Trust in 1918 to underwrite the academy in perpetuity.

In 2002 the Hershey Trust did what 99.9 percent of all corporate boards do: it decided that it would be wise to diversify its investments and that it would entertain offers to sell off its shares. The announcement sent waves of panic throughout the community as residents whose lives depended on the company contemplated the prospect of new owners gradually moving the company overseas. Local politicians, community leaders, and the unions pled with the Trust to reconsider. Pennsylvania's attorney general, Mike Fisher, went into court to stop the sale.

A TINA [there-is-no-alternative] company would have ignored the local rabble, fought the lawsuit, and kept its focus on profitability. The stakes were huge. The Chicago-based Wm. Wrigley Jr. Co. put a $12.5 billion buy-out on the table, and Nestlé and Cadbury Schweppes were reportedly prepared to offer as much as $15 billion.[2] But something miraculous happened. The Hershey Trust's board changed its mind. It reaffirmed its commitments to the community and even said that it wouldn't revisit the issue without approval from the Dauphin County Orphans Court. The Trust, of course, could change its mind again and convert the company into a TINA business (local owners always can sell out). But for the moment its decision showed how the logic governing LOIS businesses is fundamentally different from that governing TINA, fundamentally more humane, fundamentally more community-friendly.

Everything we know suggests that LOIS businesses are substantially more beneficial for a local economy than TINA businesses. This doesn't mean that TINA businesses are necessarily bad. Many sell a wonderful range of products, pay decent wages, and donate generously to local charities. But dollar for dollar of business, TINA firms contribute less to a community's well-being than LOIS firms do.

Local Is What Goes Around

The temptation, when attempting to define a qualitative word like "local" in quantitative terms, is to resort to Justice Potter Stewart's famous definition of pornography: "I know it when I see it" (*Jacobellis v. Ohio*, 1964). But let's try to do better. Perhaps the most critical element of local is proximity—both physical and geographic—because every person's purchasing choices are driven, in part, by the convenience, familiarity, and comfort of nearby stores, restaurants, professionals, and so forth.

Think of yourself as the center of your own consumption solar system, emanating rays of purchasing power. Most of your purchases are made close by—the local bank that carries the mortgage, the local clothing shops, local filling station, local charities. Travel a little outside your community and the number of purchases diminishes. Maybe it's that special trip to a mega-mall an hour away, or a consult with a medical specialist three towns over. Venture even farther out and you'll find a few purchases you make on Amazon for a book or on eBay for some rare Pokémon cards.

Each purchase you make triggers purchases by others. For instance, a dollar spent on rent might be spent again by your property owner at your local grocer, who in turn pays an employee, who then buys a movie ticket. This phenomenon is what economists call "the multiplier." The more times a dollar circulates within a defined geographic area and the faster it circulates without leaving that area, the more income, wealth, and jobs it generates. This basic concept in community economics points to the importance of maximizing the number of dollars entering a community and minimizing their subsequent departure.

The multiplier obviously diminishes with geographic distance. The farther from home you go to make a purchase; the less of the multiplier comes back and touches your community. Buy a radio down the block, the multiplier is high; buy it ten miles away, the multiplier weakens; buy it mail order, and your community gets practically no multiplier whatsoever.

There is one boundary beyond which part of the multiplier drops precipitously—that of a tax jurisdiction. A rough definition of "local," then, might be the smallest jurisdiction with real tax authority. For some this will be a town, for others it will be a city or a county. Since every purchase leads to a variety of taxes—sales taxes, wage taxes, property taxes, and business taxes—making a purchase even one village over can significantly diminish the taxes that might have gone to your own local government. For example, the savings Massachusetts consumers enjoy when they make long drives to New Hampshire malls to avoid sales taxes wind up being huge losses to the Bay State.

A business can only be considered locally owned if those who control it live in that community. That could mean the ownership is held by the sole proprietor who lives and works in the same town. It could also mean that residing in the community are more than half of a firm's partners (through a partnership, limited liability partnership, or S-corporation), shareholders (C-corporation), workers (worker cooperative or employee stock ownership plan company), or consumers (consumer cooperative). It could also refer to a nonprofit tied to the community either through its board, its mission (like a community development corporation), or through a local membership with voting rights. And it could refer to the business activities of local public agencies and public-private partnerships. There are differing consequences of each ownership structure—some, for example, are more vulnerable to a TINA takeover than others—but all offer robust benefits that stem directly from the localness of ownership and control.

There are still further complications when defining a business as local or not. Consider franchises. On paper a proprietor can own most of the outlet's capital, claim most of the profits, and yet still, by the terms of contracts and licenses, enjoy very little control. The specifics matter here. If a Subway sandwich shop is technically owned by an individual, but is largely controlled by the national chain, it cannot really be considered a LOIS business.

Or consider the residence of a proprietor in a metropolitan area. Suppose you live in Miami. Under the definition above your locality might be the city limits. Finding a local lawyer is easy. If a lawyer lives and works in Miami, she is indisputably local. But suppose she works in Miami but lives farther north in Boca Raton. Is she still local? Or what if she lives in Miami but works in Boca Raton? In either case, some of the lawyers' expenditures now leak out of Miami.

Or consider a computer purchase. You are careful to go to a locally owned electronics store but are dismayed to discover that most of the computers on display were assembled in Asia. After careful research, you finally find one model that's assembled locally and sold in the assembler's small shop. But when you crack open the machine, you realize all the components still come from Asia.

The truth is that these details matter enormously when it comes to the local multiplier. Yet few of us have time to do so much homework before every purchase. These complexities highlight why efforts to promote LOIS business involve far more than exhortations to buy or invest locally. Significant research is needed to help consumers identify goods and services with the highest degree of local content and control, and with the greatest likelihood of producing the greatest benefit for a community. The principle is easy, but its application can be difficult.

The Local Majority

Local business actually constitutes the lion's share of the U.S. economy. The U.S. Small Business Administration (SBA) defines small businesses as firms having fewer than five hundred employees, and these actually account for half of private sector employment in the country and 44 percent of private payrolls. A more restrictive definition of small business—as a firm with fewer than one hundred employees—still accounts for about a third of private employment and private payrolls. By either definition, more than 99 percent of all firms in the United States are small businesses (U.S. Census Bureau, n.d.). Put another way, footloose global businesses dominate our imagination, get showered with subsidies, and monopolize our capital markets, but actually occupy only about half of the economy. Firms with more than five hundred employees constitute only about 0.3 percent of all firms. They account for 56 percent of private payrolls, but supply fewer than half of all private jobs.

The private sector, moreover, is only part of the U.S. economy. Nearly a quarter of the nation's income, measured by the GDP, comes from household employers, nonprofits, and various government entities (U.S. Department of Commerce, n.d.). All of these categories are place based, in the sense that none of them considers setting up shop in China. Large firms turn out to be responsible for no more than 42 percent of the economy, and place-based jobs account for at least 58 percent. We can say, therefore, that *Small-Marts are responsible for most of a typical community's economy.*

This observation becomes even stronger when you consider what's left out of these tallies. Businesses with no employees, millions of which

Americans increasingly run out of their homes (many as their second or third jobs), are excluded. Another gap in the official data is unpaid household work, still done primarily by women. Were housework paid at market rates, some estimate this additional income would account for as much as a quarter of the economy (Cahn 2003, 1001–04).[3] If the volunteers serving senior citizens were paid a wage of eight dollars per hour, the total value of the services would exceed the actual cost of formal home health care and nursing home care (Cahn 2003, 1001–04; Levine & Memmott, n.d., 182–88). The value of all volunteer efforts in the country, of course, is greater still. Another gap, according to Edgar Feige of the University of Wisconsin, is the underground or black economy, mostly homegrown, valued somewhere between $500 billion and $1 trillion ("The Underground Economy," 1998). As law professor Edgar S. Cahn (2003, 1001–04) writes, "A wide range of estimates—from Gary Beck to Nancy Folbre—finds that at least 40% of our country's productive work goes on outside of the market economy." Nearly all of these missing pieces are local, which means that a better accounting system would most likely show that *at least three-quarters of our economic activity is currently place based.*

We must also remember that the Small-Mart sector is unevenly distributed throughout the country; in some regions its participation is significantly higher than the overall average. In a quarter of the states, firms with more than five hundred employees account for less than half of private payrolls. Move into suburban and rural areas, and the role of small businesses gets larger still. In Montana and Wyoming only four out of ten payroll dollars come from large firms, suggesting that the Small-Mart portion of their economies, in conventional accounting terms, probably hovers around 70 percent.

But don't small businesses represent the backwater of business, the inefficient remnants of the old economy? Hardly. According to the SBA, small firms generate 60 to 80 percent of all new jobs and produce thirteen to fourteen times more patents per employee than large firms.

What about the high failure rate of small business? The SBA reports that a third of small business start-ups shut down within two years, and half within four years. These figures are sometimes tossed around to suggest how unreliable Small-Marts are, but the real story is much more interesting. The failure rates only refer to start-ups, not existing small businesses. Owners of a third of the closures, moreover, actually pronounce their ventures successful (for example, an entrepreneur who operates a home-based catering business for a few years that then serves as a launching pad for a new restaurant) (Headd 2003, 51–61). And here's another surprising fact: for almost every one of the last ten years the birth rate of small businesses has exceeded the death rate, while for large firms the death rate has been greater than the birth rate. Between 2000 and 2001 for example, 553,000 small businesses closed but 585,000 opened, with a net increase of 32,000 firms (U.S. Census Bureau, n.d., 495).[4] The total universe of existing small business was about 5.6 million firms, which means that in any given year, about 90 percent of existing small businesses continue to compete effectively. During the same 2000 to 2001 period there was a net loss of about two hundred large firms.

How has globalization changed the role of small business? Over the past decade, while globalization was becoming a household word, a shift in favor of larger businesses has occurred, but arguably only a modest one. Between 1990 and 2001 about 4 percent of the jobs shifted from very small to large firms. Firms with 100 to 499 employees remained steady.

These data lend themselves to several different interpretations. On the one hand, you could conclude that the mighty gales of global "creative destruction," in the famous phrase of economist Joseph Schumpeter, caused surprisingly little change in the composition of the economy. After all, fewer than four out of a hundred workers were affected, and more than 96 percent of the size structure of the economy remained stable (and again, a shift that is even smaller when one considers home-based, non-employee, and unpaid workers). On the other hand, you could see this as proof that TINA-style globalization has taken a serious toll on the smallest businesses in the United States and that if the next several decades look like the previous decade, small businesses could become as rare as the spotted owl.

Whichever view you choose, the question of why TINA did slightly better than LOIS during those years remains. Most economists would say that these trends prove the greater efficiency and superior performance of TINA firms. Global companies, taking advantage of economies of larger scale as well as lower wages and looser environmental standards abroad, are now producing the cheapest goods and supplying the most cost-effective services, undercutting an increasing number of local businesses. This interpretation omits, however, myriad "imperfections" in our market economy that uniformly favor TINA.

Consider two other important stories about the relative strength of LOIS versus TINA businesses that are mutually contradictory. . . . One story is of massive consolidation. Between 1998 and 2002 the sector that experienced the greatest degree of consolidation was the securities industry, reflecting the decision of Congress to remove the regulatory barrier between banking and securities.[5] Broadcasting and telecommunications, another industry undergoing massive deregulation, was second. Hospitals have also become more centralized—a direct result of the health care crisis—and the growth of chain stores has contributed to the consolidation of retailing of clothing, electronics, and sporting goods. Each consolidating industry could receive a dissertation's worth of scrutiny of the people, technology, innovations, and laws responsible for the shift in scale.

But an equally important story is that almost as many sectors in the economy are actually decentralizing. Investment advising for trusts and estates has gone local. Minimills for steelmaking are doing well. Utilities are shrinking in size. Even as some textile, clothing, and transportation equipment manufacturing moves overseas, smaller plants in these sectors are expanding.

Which story is right? Well, both are. And they suggest that for every piece of bad news for Small-Marts, there's good news as well. In many sectors Small-Marts are innovating, taking advantage of cutting-edge ideas in marketing and technology, and making inroads against larger business. And looking ahead . . . , the most important trends in the global economy actually favor the expansion of Small-Marts. That doesn't mean that the Small-Mart Revolution is inevitable, especially if the current pro-TINA biases in subsidies, capital markets, and economic-development practice are not undone. The Small-Mart Revolution requires dramatic changes in the behavior of consumers, investors, entrepreneurs, and policymakers.

But let's return to a more basic question: Why should we favor LOIS and join the revolution at all?

Swing LO, Sweet Business

However we define local ownership, it turns out to be an essential condition for community prosperity for at least five reasons, spelled out below. The first four flow from the inherent difference between LOIS and TINA firms: most local entrepreneurs form their businesses in a particular place because they love living there. For a few sophisticated LOIS entrepreneurs, other factors like taxes, workforce quality, and clusters may come into play, at least in their initial decision about where to set up the business. But once a LOIS enterprise is up and running, the entrepreneur's family, workers, and customers are woven into the fabric of the community, and as a result, he or she has relatively little interest in moving to Mexico or Malaysia.

Local Ownership Advantage #1: Long-Term Wealth Generators. Because their entrepreneurs stay put, LOIS businesses are more likely than TINA ones to be cash cows for communities for many years, often for many generations. The Hershey Chocolate Company has brought tens of billions of dollars into the community over its lifetime and will do so for the foreseeable future.

Local Ownership Advantage #2: Fewer Destructive Exits. The anchoring of LOIS businesses minimizes the incidence of sudden, calamitous, and costly departures. For about a century, the economy in Millinocket, Maine, was built around the Great Northern Paper Company, which was one of the largest and most advanced paper manufacturers in the world. During Christmas of 2002 the owners of the company living "away" decided that operations would be better based elsewhere, and the last fourteen hundred workers were laid off. For the next year the unemployment rate hovered at about 35 percent, higher than what the country endured during the Great Depression. This kind of death spiral—a sudden departure followed by massive unemployment, shrinking property values, lower tax collections, deep cuts in schools, police, and other services, which throws still more people out of work, and so forth—is far less likely in a regional economy made up primarily of LOIS businesses.

Local Ownership Advantage #3: Higher Labor and Environmental Standards. A community made up mostly of LOIS businesses can better shape its laws, regulations, and business incentives to protect the local quality of life. A TINA-dependent community is effectively held hostage to its largest TINA companies. While not shy about lobbying politicians, locally owned companies usually do not threaten to leave town. A community filled primarily with locally owned businesses can set reasonable labor and environmental standards with confidence that these enterprises are likely to adapt rather than flee. For example, on Maryland's Eastern Shore, two powerful poultry companies, Tyson and Perdue, have successfully fought legislative efforts to raise their workers' wages or clean up the billions of pounds of chicken manure they dump into the Chesapeake Bay ecosystem by deploying one powerful argument: regulate us and we'll move to more lax jurisdictions like Georgia or Arkansas. . . .

Local Ownership Advantage #4: Better Chances of Success. In November 2003 I debated Jack Roberts, the head of Lane County Metro Partnership, the principal economic development organization in the region surrounding Eugene, Oregon. Like other developers in the state, Roberts handed out tax abatements to businesses as an incentive to either move to the area or expand. . . . [A]bout 95 percent of his abatements were used to lure six TINA companies to move in, while the other 5 percent were given to dozens of LOIS businesses. Ultimately, according to an investigative report in the local newspaper, the cost to the community in lost taxes was about $23,800 per job for the TINA firms and $2,100 per job for the LOIS firms (McDonald & Wihtol 2003; Shuman 2004).[6] Why were the TINA jobs *more than ten times* more expensive? Roberts argued that it was just bad luck. The firms recruited were mostly high tech, and when investors lost faith in the tech sector around 2000 and 2001, management cut costs by shutting down the plants. What Roberts' argument overlooks, however, is that business cycles are *always* oscillating, and during inevitable down periods a TINA business will be prone to consider moving a factory to a lower cost region. In fact, even during up periods, a TINA business will consider moving if the rate of return on investment can be ratcheted a notch or two higher. Why keep a factory open in Eugene, earning a 10 percent return, if it can earn 20 percent in Bangalore? To a LOIS entrepreneur, in contrast, these community-destroying options are off the table.

Local Ownership Advantage #5: Higher Economic Multipliers. In the summer of 2003, a consulting group of economists called Civic Economics studied the impact of a proposed Borders bookstore in Austin, Texas, compared with two local bookstores.[7] They found that one hundred dollars spent at the Borders would circulate thirteen dollars in the Austin economy, while the same one hundred dollars spent at the two local bookstores would circulate forty-five dollars—roughly three times the multiplier. In 2004 Civic Economics completed another study of Andersonville, a neighborhood in Chicago. The principal finding was that a dollar spent at a local restaurant had 25 percent more economic impact than a chain. The local advantage was 63 percent more for local retail, and 90 percent more for local services.

This last point, largely unfamiliar to economic developers, is worth further elaboration. A study of eight local businesses in the towns of Rockland, Camden, and Belfast [Maine] found that they spent 45 percent of their revenue within their local counties, and another 9 percent statewide. The aggregate level of in-state spending was nearly four times greater than that from a typical chain store. Other studies in the United States and abroad also have found that local businesses yield two to four times the multiplier benefit as comparable nonlocal businesses.[8]

Skeptics complain that these multiplier studies are flawed. They claim that the economic models used are filled with uncertainties, especially when a small locale is under study; that the multipliers studied are always specific to a location, so generalizations are difficult; and that the TINA and LOIS businesses being compared are really so different that it's like comparing apples to goldfish. They also contend that the results are unreliable because the researchers have only partial information on the chain firms' expenditures.

Whatever the merit to these objections, the skeptics overlook three points. First, the authors of these studies, unlike some of their TINA scholar counterparts, are honest enough to point out the flaws in their own methodology. Second, these flaws differ little from the flaws of the pro-TINA studies, like the Moore School's puff piece on South Carolina's investment in BMW. Third and most important, the underlying reason why local businesses have higher multipliers is obvious and unlikely ever to change: *they spend more locally*. In the Austin analysis, local bookstores, unlike Borders, have local management, use local business services, advertise locally, and enjoy profits locally. These four items alone can easily constitute a third or more of a business's total expenditures. That LOIS businesses almost always spend more locally means that they almost always yield a higher multiplier.

To recap all these advantages, look at the National Football League. All but one franchise is owned by a single (usually obnoxious) individual, and these modern moguls have threatened to split town if demands for hundreds of millions of dollars for new stadiums and salary increases are not met. When Cleveland refused, Art Modell, owner of the Browns, took *his* team to Baltimore. The one exceptional franchise is the Green Bay Packers, a community-controlled nonprofit, whose shareholder-members are primarily citizens of Wisconsin.[9] Because its fans will never allow the team to leave town, the Packers have become a critical source of wealth and economic multipliers for Green Bay, one that will be around for generations of Cheesehead fans to come. Being locally controlled means the team cannot suddenly depart and punch a hole in the economy, even if its rate of return might be higher somewhere else. If the city ever passed a living wage ordinance, the Packers would learn to adapt, since fleeing is not an option.

If local ownership of a football team can confer all these benefits, doesn't it make sense to insist on local ownership of farms, factories, and banks?

What the Meaning of IS [Import Substitution] Is

Here's a quick recipe for local prosperity: create a diversity of locally owned businesses, design them to use local resources sustainably, and make sure that together they are fully employing residents and producing at least enough goods and services to satisfy residents' needs. For those needs that cannot be met through local production, export enough goods and services to provide residents with the income to buy needed imports. To understand this formula, consider two ways how *not* to create a thriving local economy.

Suppose your community were completely self-reliant. You and your Robinson Crusoe brethren might build houses out of local wood and stone, grow food in the community greenhouses, draw water from rooftop rain collectors, and so forth. This kind of primitive economy can work, provided you're willing to forgo all the products and technology originating from elsewhere on the planet. For most us this is inconceivable. Even those of us who embrace lives of "voluntary simplicity" have many clothes on our back, couches in our living rooms, and computers on our desks that come from elsewhere. We need to sell something to buy these goods. Absolute self-reliance won't work.

Next, suppose your community met only other people's needs—that is, your businesses were 100 percent dedicated to exports, like the Mexican maquiladoras that line the southern border of the United States. Now you're making money, but you're also a sitting duck that can be blown away by outsiders' economic shotguns. If you followed economists' advice to find your special niche in the global economy, you might be exporting one or two products, and your well-being would be totally dependent on the stability of those global markets. Lose your lead, as Detroit did with automobiles and Youngstown did with steel, and your economy collapses. Plus, you're vulnerable to all kinds of nasty surprises because you're importing everything else, even your most basic needs, which now must come in the form of canned food and bottled water. The best example of this is the U.S. economy's dependence on foreign oil, which ties us—like a damsel in a bad melodrama—to sudden OPEC-orchestrated spikes in global petroleum prices and requires a foreign policy weighed down by increasingly expensive and bloody military involvements in the Middle East.

A better alternative is to blend the two extremes. The healthiest economy is both self-reliant *and* a strong exporter. Meet as many of your own needs as possible, *then* compete globally with a diversity of products. By being relatively self-reliant, you're far less vulnerable to events outside your control. By having global sales, you're not closing off your economy to outside goods and technology. Meanwhile, you're conducting as much business as possible with both local and foreign consumers, which bring wealth into the community and pumps up the multiplier. Cut back on *either* self-reliance or exports, and you lose income, wealth, and jobs.

This may seem contradictory. If every community in the world became more self-reliant, wouldn't the aggregate level of imports shrink and make it difficult, if not impossible, for communities to increase their exports? In the short term, yes. But over the long term, import substitution would enable tens of thousands of communities worldwide to stop wasting precious earnings from exports on imports they could just as easily produce for themselves, and encourage them instead to reinvest those earnings in industries that could truly fill unique niches in the global economy. It's a mistake to view any economy, especially the world's, as a zero-sum game where one player's gain is another's loss.

The relationship between any two communities in the global economy is not unlike a marriage. As couples counselors advise, relationships falter when two partners are too interdependent. When any stress affecting one partner—the loss of a job, an illness, a bad-hair day—brings down the other, the couple suffers. A much healthier relationship is grounded in the relative strength of each partner, who each should have his or her own interests, hobbies, friends, and professional identity, so that when anything goes wrong, the couple can support one another from a position of strength. Our ability to love, like our ability to produce, must be grounded in our own security. And our economy, like our love, when it comes from a place of community, can grow without limit.[10]

If it's important to develop strong exports and to be self-reliant through import substitution, should both strategies be implemented simultaneously or should one be prioritized over the other? The prevailing view among state and local economic development experts is to prioritize exports. That's why they spend millions of dollars to lure and keep TINA businesses. Only through export earnings, as the Moore School scholars argued about South Carolina's decision to give millions to BMW . . . , can a community enjoy the potentially unlimited fruits of *new* dollars.

But the argument is flawed. How does a dollar brought into the community from export sales differ from a dollar retained in the community's economy through local sales? From a multiplier standpoint, there's no difference whatsoever. One academic analysis of eight southeastern states, looking at the relationship between local services and nonlocal non-service industries like manufacturing and mining, found both dimensions of the economy equally important. After reviewing this data, Thomas Michael Power, chair of the economics department at the University of Montana, observes: "Growth in service activities played a very important role in determining overall local economic growth. Manufacturing and other export-oriented activities were not the primary economic forces. Others have also found evidence that 'local' economic activities may drive the overall economy rather than just adjust passively to export activities" (Power 1996, 125).

Even though development through import replacement and development through exports propel one another, there are many compelling reasons to favor the former from a public policy standpoint. Import substitution involves shifting purchases from businesses outside the community to those inside, which usually means from businesses owned by outsiders to those owned locally. All the benefits of local ownership are therefore reinforced through import substitution. Every time a community chooses to produce its own apples rather than import them, assuming that the prices of all apples are roughly equal, it boosts the economic well-being of its own apple farmers, as well as all the local suppliers to the farmers and all the other local businesses where the farmers spend their money.

Export-led development means opening yourself up to many otherwise avoidable dangers. Importing oil leaves the fate of our economy in the hands of OPEC ministers, Latin American strongmen, and Arab sheikdoms. Importing Canadian beef invites outbreaks of mad cow disease. Importing chickens puts you on the front lines of the avian flu pandemic. . . .

Supporting the development of diverse enterprises—enacting import-substitution policies—enhances the skill base of a community and acts as a kind of insurance policy, an investment in the people, know-how, and technology that can enable you to take full advantage of the "next big thing" in the economy. A generation ago a Boeing-dependent Seattle could not have possibly known that its future lay, not in aerospace, but in software and coffee. You never can know, and it is only by having a diversified economy, as Seattle did, that you can have the skills to seize whatever opportunities arise.

Paradoxically, import substitution also turns out to be the best way to create a healthy export sector. An unhealthy approach to exports is to do what Millinocket, Maine, did, which, as noted earlier in this chapter, put all its economic eggs in the basket of paper production. Similarly, when economic developers attempt to divine what your community's one or two great "niches" might be in the global economy, they are essentially playing a dangerous game of Russian roulette. If your niche suddenly becomes obsolete, you're dead. A far smarter approach is to invest in dozens of local small businesses, all grounded in local markets, knowing that some will then develop a variety of healthy export markets. A multiplicity of export linkages is the most powerful and safest way to compete globally.

Suppose North Dakota wished to replace imports of electricity with local wind-electricity generators. Once it built windmills and became self-reliant on electricity, it would then be dependent on outside supplies of windmills. If it set up a windmill industry, it would become dependent on outside supplies of machine parts and metal. This process of substitution never ends. But it leaves North Dakota with many strengthened local industries—in electricity, windmills, machine parts, and metal industries—that not only can meet local needs but also can take advantage of export opportunities.

Even if import replacement leads to more exports, the distinction between this process and export-led development is much more than simply a matter of semantics. Had South Carolina followed an import replacing development strategy, it would have used the same money it paid BMW—or much less—to nurture hundreds of existing, locally owned businesses, some of which would have then become strong exporters. Development led by import replacement rather than export promotion diversifies, stabilizes, and strengthens the local economy, while allowing the best exporters to rise on their own merits. As Thomas Michael Power (1996) says, "Export-oriented economies remain primitive, suffer through booms and busts, and go nowhere. It is only when an area begins making for itself what it once imported that a viable economic base begins to grow."

This touches on a final advantage of import-substituting development. Which is easier: for governors, mayors, and economic developers to learn a foreign language like Japanese, travel abroad to snag some new global company, steal tens of millions of dollars from the taxpayers to provide the necessary incentives, and then have to defend the decision a decade later when the company moves on; or for the same folks to speak in plain English with their own business community, work together on nurturing homegrown enterprises, and enjoy the fruits of their efforts when they retire? The excitement officials and civil servants feel when they travel to exotic lands and rack up the frequent flier miles is understandable, but it should never be done on community time. . . .

Not every LOIS business is a model environmental citizen—one can certainly point to small-scale manufacturers and local dry cleaners that release carcinogens—but an economy made up largely of LOIS business is more likely to be green. Local ownership provides an important form of ecological accountability since the owner must breathe the same air and drink the same water, and his or her family must ultimately live side by side with the rest of the community. Moreover, many LOIS businesses are service related, and these usually are labor intensive and have fewer environmental impacts. As noted earlier, a community with primarily locally owned businesses—businesses that will not consider moving to Mexico or China—can raise environmental standards with greater confidence that these firms will adapt, a circumstance that tips the political balance in favor of tougher environmental regulations.

A TINA-dependent community, in contrast, is likely to suffer several kinds of environmental hazards. Box stores, for example, are characterized by gigantic parking lots, which cover vast tracts of land with concrete that drain off oil, gasoline, and other toxins into the water table, often in torrents that can lead to flooding. When national chains move on, these huge spaces are neglected, become eyesores, and lower property values. Nationwide Wal-Mart has three hundred vacant stores, and most are less than a mile away from the Supercenter that took the predecessor store's place (Mitchell 2004, 9).

The relative immobility of LOIS businesses also serves the rights of labor, though this argument contradicts the historic hostilities union organizers and old-school lefties harbor toward small business. Their concern has been that, compared to larger businesses, small businesses pay lower wages, provide fewer benefits, and are less susceptible to union organizing. There is evidence, to be sure, that businesses with more than five hundred employees pay about a third more on average than businesses with fewer than five hundred employees. But one recent statistical analysis of the relevant academic literature found that between 1988 and 2003 these differences, in both wages and benefits, shrank by about a third (Hollister 2004, 659–76). If this trend continues—especially as many of the once high-paying larger firms continue to move factories overseas and as low-wage retailers like Wal-Mart continue to displace existing small business—these differences could disappear altogether.

TINA businesses that once offered fabulous worker benefits are now chopping them away, as more and more managers struggle to contain ballooning health care costs and place responsibility for pension contributions directly on the employee. The growing incidence of TINA firms declaring bankruptcy (including United Airlines, a company controlled by its supposedly enlightened workforce) as a strategy to escape long-standing health plans and pension benefits should give pause to anyone who thinks that big business is the ticket to economic security. The real solutions for all Americans to have better health care and retirement—and not just those employed or employable—must come, as they do in almost every other industrialized country, from smarter public policy. . . . In fact, public policies that do a better job of ensuring these benefits for all workers may eliminate one of the big reasons some choose to work for TINA firms and expand the number and quality of people eager to work in small business.

Small businesses may be less easy to unionize than large ones, but that doesn't necessarily make them less sensitive to labor rights. Some of the most socially responsible entrepreneurs in this country are the small business pioneers who are members of organizations like BALLE and SVN and who believe that high wages and decent benefits are not just good motivators but also moral imperatives. The closeness of the relationships between the people on the top and the bottom of these small firms also can be a powerful force for empathetic management. And it seems ludicrous for labor to favor TINA businesses when nearly all of them, by now, cannot wait to purge their businesses of unions by moving production overseas.

Sooner or later, the labor movement in the United States will recognize that TINA enterprises have become dead ends for vindicating the rights of workers. Labor should embrace small business, unionize it where it can, and encourage worker ownership, participation, and entrepreneurship where it can't. Meanwhile, higher community standards through living wages . . . and serious health care reform are probably the most effective ways of helping all workers, irrespective of the size of their employer.

Another sign of a prosperous community is how well it preserves its unique culture, foods, ecology, architecture, history, music, and art. LOIS businesses celebrate these features, while chain stores steamroll them with retail monocultures. Austin's small business network employs the slogan "Keep Austin Weird." Outsider-owned firms take what they can from local assets and move on. It's the homegrown entrepreneurs whose time horizon extends even beyond their grandchildren and who care most about preserving these assets. And it's the local marketers who are most inclined to serve local tastes with specific microbrews and clothing lines. "Weirdness" is what attracts tourists, engages locals in their culture, draws talented newcomers, and keeps young people hanging around. As Jim Hightower (2004) writes, "Why stay at the anywhere-and-nowhere Holiday Inn when we've got the funkily refurbished Austin Motel right downtown, boasting this reassuring slogan on its marquee: 'No additives, No preservatives, Corporate-free since 1938.'" . . .

What about a community's social well-being and political culture? In 1946 two noted social scientists, C. Wright Mills and Melville Ulmer, explored this question by comparing communities dominated by one or two large manufacturers versus those with many small businesses. They found that small business communities "provided for their residents a considerably more balanced economic life than did big business cities" and that "the general level of civic welfare was appreciably higher" (Lyson 2001, 3). A congressional committee published the study, and in the foreword, Senator James E. Murray wrote:

> It appears that in the small-business cities is found the most favorable environment for the development and growth of civic spirit. A more balanced economic life and greater industrial stability is provided in the small-business cities. There the employment is more diversified, the home-owning middle class is larger, and self-employment greater. Public health is greater . . . the study reveals that a baby has a considerably greater chance to survive in his first year in the small-business city than in the one dominated by a few large firms (Lyson 2001, 12–13).

Thomas Lyson, a professor of rural sociology at Cornell University, updated this study by looking at 226 manufacturing-dependent counties in the United States. He concluded that these communities are "vulnerable to greater inequality, lower levels of welfare, and increased rates of social disruption than localities where the economy is more diversified" (Lyson 2001, 14).

We know that the longer residents live in a community, the more likely they are to vote, and that economically diverse communities have higher participation rates in local politics. Moreover, Harvard political scientist Robert Putnam has identified the long-term relationships in stable communities as facilitating the kinds of civic institutions—schools, churches, charities, fraternal leagues, business clubs—that are essential for economic success. As one group of scholars recently concluded after reviewing the social science literature: "The degree to which the economic underpinnings of local communities can be stabilized—or not—will be inextricably linked with the quality of American democracy in the coming century" (Williamson, Imbroscio, & Alperovitz 2003, 8). A LOIS economy with many long-term homegrown businesses is more likely to contribute to such stability than the boom-and-bust economy created by place-hopping corporations.

But perhaps the most important benefit of spreading LOIS businesses is that it allows a community to rehumanize the economic relationships among its residents and reassert control over its destiny. . . .

The Challenges of Social Responsibility

. . . Social responsibility must include local ownership. . . . It's so disheartening to see a proliferation of nonprofits, books, green directories, conferences, and declarations proclaiming social responsibility without ever a mention of the issues of ownership or control. In early 2003 California State Senator Alarcon introduced a bill in the California state legislature (SB 974) that would have awarded 5 to 10 percent bidding preferences on state and local government contracts whenever a business achieved ten of thirteen criteria for social responsibility. It duly recognized corporations paying living wages, providing health insurance and retirement plans, promoting recycling, implementing job retention, and respecting consumer safety. But what about ownership? Except for a vague criterion encouraging "worker involvement or worker ownership," there was no mention of *local* ownership whatsoever. Yet it also needs to be said that the criteria of the Alarcon bill are important and that they do not automatically flow from local ownership. No company structure, on its own, can guarantee that managers always do the right thing. Corporate responsibility requires LOIS but also two kinds of supplements.

First, a prosperous community requires healthy local governance so that reasonably high labor and environmental standards are set for all business. The "High Road" in economic development, as Dan Swinney, a sharp Chicago-based organizer, calls it, inevitably demands that public bodies set speed limits, rights of way, and traffic signals for commerce (Swinney 1998). For example, enacting a living wage ordinance as did the city of Santa Fe, New Mexico, raised labor standards significantly (Gertner 2006). An economy made up mostly of LOIS businesses may make it politically possible to enact a living wage, though it does not follow that LOIS businesses automatically will embrace it (many didn't in Santa Fe). They will, however, adapt to it over time, because moving is not an option and shutting down is not in their interest.

A second mechanism is to nurture more enlightened shareholders. Some owners of LOIS businesses—family members, partners, friends, colleagues, and other investors—can be just as brutal in demanding that managers pay attention to the short-term bottom line as the faceless stockholders of publicly traded companies. A healthy LOIS economy ultimately requires activist shareholders who are capable of balancing the interests of the company with those of the community so that when a living wage ordinance is passed, they don't react by shutting down.[11] Because the shareholders live in the community and presumably know, appreciate, and even honor many of their neighbors, they are more likely than absentee owners to make more community friendly choices—but they won't do so automatically. Public education and peer pressure must remind shareholders that their responsibility is to discharge their duties to *both* the company and the community in a balanced way.

The private and public spheres of a community are intimately related, and the tone and activities of one influence the other. An economy comprising mostly LOIS enterprises can weave together peer relationships among businesses, and between businesses and others, that facilitate communication, discourse, reason, even empathy, all of which are necessary for good governance and high stakeholder awareness.

Many economists concede that, *in theory*, a community rich with LOIS businesses will prosper. Yes, a community made up of locally owned businesses will enjoy more engines of wealth, over many more years, with less worry about catastrophic departures, and with greater multipliers for every dollar of business. And a self-reliant community will be more secure, better able to tap a deeper pool of labor skills, benefit from a wider range of connections to the global economy, and celebrate that its economic development programs, now shorn of outrageous incentives and extravagant junkets, are cheaper and more cost effective. *But*, insist these dismal social scientists, we are in an era where bigger is better. The most competitive goods and services can only come from larger TINA firms, and the consumer advantages they confer outweigh any potential community advantages from LOIS firms.

Were this dilemma real, if we had to choose between competitive goods and services from community-destroying TINA firms and uncompetitive goods and services from community-friendly LOIS firms, picking the right future would be agonizingly difficult. Fortunately, LOIS firms are far more competitive than almost anyone realizes.

Notes

1. See Hoover's database on "Companies and Industries," www.hoovers.com. See also www.hersheytrust.com and Michael D'Antonio, *Milton S. Hershey's Extraordinary Life of Wealth, Empire, and Utopian Dreams* (New York: Simon & Schuster, 2006).

2. "Wrigley Challenges Confectionery Giants," *Confectionery News*, 14 September 2005; "Hershey Trust Bows to Pressure over Future Sale," *Confectionary News*, 27 September 2002; and "Hershey Kisses Company Sale Goodbye," 18 September 2002, www.isa.org.

3. Cahn quotes, for example, estimates made by the think tank Redefining Progress.

4. It is worth pointing out the limits of these data: "Although the data sources mentioned . . . put great effort into finding new firms promptly, determining when new firms close, dealing with unreported data, and identifying mergers and spin-offs, they still imperfectly represent the universe of firms that is their target." Catherine Armington, "Development of Business Data: Tracking Firm Counts, Growth, and Turnover by Size of Firms," monograph (SBA Office of Advocacy, Washington, DC, December 2004): 34.

5. The choices of years for comparison may seem arbitrary, but they reflect: (a) that federal accounting of sectors made a major change in 1998, from Standard Industrial Codes (SIC) to the North American Industry Classification System (NAICS); and (b) that 2002 is the most recent data available.

6. The calculation per TINA job is based on the gross number of total jobs the six companies were providing in mid-2003. Were the measurement done on the basis of net increases in jobs after the subsidies, the cost per TINA job would be $67,220—or thirty-three times greater than the cost per LOIS job.

7. The Austin study is "Economic Impact Analysis: A Case Study," monograph (Civic Economics, Austin, Texas, December 2002). The Andersonville study is "The Andersonville Study of Retail Economics" (Civic Economics, Austin, Texas, October 2004). Both can be downloaded free of charge at www.civiceconomics.com. "The Economic Impact of Locally Owned Businesses vs. Chains: A Case Study in Midcoast Maine," monograph (Institute for Local Self-Reliance and Friends of Midcoast Maine, September 2003).

8. See, for example: David Morris, *The New City-States* (Washington, DC: Institute for Local Self-Reliance, 1982), 6 (showing that two-thirds of McDonald's revenues leak out of a community); Christopher Gunn and Hazel Dayton Gunn, *Reclaiming Capital: Democratic Initiatives and Community Control* (Ithaca, NY: Cornell University Press, 1991) (finding that 77 percent of a typical McDonald's "social surplus" leaves the community); Gbenga Ajilore, "Toledo-Lucas County Merchant Study," monograph (Urban Affairs Center, Toledo, OH,

21 June 2004) (calculating an economic impact of a local bookstore more than four times greater than that of a typical Barnes & Noble); Justin Sachs, *The Money Trail* (London: New Economics Foundation, 2002) (spelling out a multiplier methodology used by communities throughout the United Kingdom, and documenting case studies showing how local businesses double or triple the economic impact of nonlocal competitors).

9. For a more extensive discussion of the Packers, see Michael H. Shuman. *Going Local: Creating Self-Reliant Communities in a Global Age* (New York: Routledge, 2000), 3–6.

10. This does not mean *conventional* economic growth can continue ad infinitum, because ultimately the world will run up against finite energy and mineral resources and limits to how much waste its ecological sinks can absorb. It is possible, however, to imagine the infinite growth of ingenuity if it's also accompanied by reductions in the consumption of physical resources and energy.

11. See, for example, Bruce T. Herbert's insistence on asking Weyerhaeuser tough questions, forcing the board to rescind its no Q&A policy at a recent shareholder meeting. Gretchen Morgenson, "Managers to Owners: Shut Up," *New York Times*, 24 April 2005. For examples of creative institutional shareholder activism, see Marjorie Kelly, David Smith, and Nicholas Greenberg, "Transforming Economic Power: State and Local Approaches to Corporate Reform," monograph (New York: Demos, 2005): 8–11.

References

Cahn, Edgar S. 2003. "Nonmonetary Economy." In *Encyclopedia of Community*, eds. Karen Christensen and David Levison, 1001–04. Thousand Oaks, CA: Sage.

Gertner, Jon. 2006. "What Is a Living Wage?" *New York Times Magazine*. (January 15).

Halweil, Brian. 2002. *Home Grown: The Case for Local Food in a Global Market*. Worldwatch Paper no. 163. Washington, DC: Worldwatch.

Headd, Brian. 2003. "Redefining Business Success: Distinguishing Between Closure and Failure." *Small Business Economics* 21: 51–61.

Hightower, Jim. 2004. "Whose Town Is It?" *The Austin Chronicle* 20 (February).

Hollister, Matissa N. 2004. "Does Firm Size Matter Anymore? The New Economy and Firm Size Wage Effects." *American Sociological Review* 69 (October): 659–76.

Jacobellis v. Ohio, 378 U.S. 184, 197 (1964).

Levine, Arno C. and Memmott, M.M. "The Economic Value of Informal Caregiving." *Health Affairs* 18 (2): 182–88.

Lovins, Amory B., and Lovins, L. H. 1982. *Brittle Power: Energy Strategy for National Security*. Andover, MA: Brickhouse.

Lyson, Thomas A. 2001. "Big Business and Community Welfare: Revisiting a Classic Study." Monograph (Cornell University Department of Rural Sociology, Ithaca, NY): 3.

McDonald, Sherri Burri, and Wihtol, Christian. 2003. "Small Businesses: The Success Story." *The Register-Guard* (August 10).

Mitchell, Stacy. 2004. "10 Reasons Why Maine's Homegrown Economy Matters: And 50 Proven Ways to Revive It." Monograph (Maine Businesses for Social Responsibility, Belfast, ME, June): 9.

Power, Thomas M. 1996. *Environmental Protection and Economic Well-Being*. Armonk, NY: M.A. Sharpe.

Shuman, Michael H. 2004. "Go Local and Prosper." *Eugene Weekly* (January 8).

Swinney, Dan. 1998. "Building the Bridge to the High Road." Monograph (Midwest Center for Labor Research, Chicago, Illinois).

"The Underground Economy." Brief Analysis no. 273. (National Center for Policy Analysis, 13 July 1998).

U.S. Census Bureau. *Statistical Abstract of the United States: 2004–2005*. Table No. 732, 494.

U.S. Census Bureau. *Statistical Abstract*. Tables No. 733–34, 495.

U.S. Department of Commerce. Bureau of Economic Analysis. Table I.3.5 on "Gross Value Added by Sector." Available at http://www.bea.gov/national/nipaweb/index.asp.

Wessell, David. 2001. "Capital: Decentralization and Downtowns." *Wall Street Journal*. (October 25).

Williamson, Thad, David Imbroscio, and Gar Alperovitz. 2003. *Making a Place for Community: Local Democracy in a Global Era*. New York: Routledge.

8

Daring to Experiment*

Warren A. Johnson

When economic growth was becoming the root metaphor and dominant principle of society, the prior cultural emphasis on stability must have seemed like an anchor holding back progress. Now, driven by deep social and ecological concerns, the desire for stability has reappeared. Here geographer Warren A. Johnson dares to consider a much-maligned social order of the past—feudalism. In this work, he asks whether a carefully reconstituted form of feudalism, one based on fair contracts negotiated with landowners and entered into freely by community members, may actually have advantages in the transition toward durable living, including the benefit of social stability.

Feudalism is, of course, far from fashionable these days, and deservedly so given its history of fiercely suppressed personal freedom. Certainly localization will never flourish if it involves a return to such an unjust order. Yet Johnson argues that feudalism deserves an honest appraisal. It did help protect a rich European rural landscape while supporting substantial numbers of people. Perhaps there are lessons we can extract from this pre–fossil fuel era for the coming post–fossil fuel one, especially if we incorporate other hard-learned lessons concerning democracy, human rights, and well-regulated markets from the last several centuries (see part V). Notice that access to land through alternative ownership structures might be one such idea to emerge from a frank consideration of feudalism.

During the coming transition, and despite discomfort with ideas like feudalism, localizers will need to conduct multiple experiments to find

*Johnson, Warren A. 1972. "Paths out of the corner," *IDOC International,* North American edition, 47. Excerpted and reprinted with permission. See also his *Muddling toward Frugality* (1978, Sierra Club Books) and *The Gift of Peaceful Genes* (2002).

Return of the Erie Canal

Mention the Erie Canal and most Americans' minds fill with stories of America's pioneering past. According to Craig Williams, history curator at the New York State Museum in Albany, the canal reached its peak as a transporter of goods and people in 1855. Canal traffic declined dramatically in the late 1800s but the canal persevered until the 1950s, when interstate highways and the St. Lawrence Seaway almost killed it. For many, the canal became just a memory, although a few found it to be a scenic place for recreational boating. Then, in the 2000s, diesel fuel and shipping costs began to rise and a few shippers rediscovered the Erie Canal. They found the historic canal to be a fuel-efficient method for shipping goods between the upper Midwest and the East Coast. In commenting on the return of commercial shipping, Carmella R. Mantello, director of the New York State Canal Corporation, which operates the Erie Canal, said in 2008, "We anticipated we might have an increase in commercial traffic, but nowhere near what we're seeing today."

In the short term, if fuel prices dropped, canal traffic would likely diminish again. But long-term trends in fossil fuel availability suggest water transport may once again be a wise choice. In fact, the current traffic has caught the attention of business owners located nearby. David Colegrove, president of Auburn Biodiesel, located adjacent to the canal, took note of the growth in shipments on the canal. He now hopes to bring soybeans in by canal barge and ship finished products to New York City. "The amount of money you can save is really eye-popping," he said; "I'm fascinated by the history of the canal, and I'm intrigued by how well it still works."

Sources: Christopher Maag, "Hints of Comeback for Nation's First Superhighway," *New York Times*, November 3, 2008.

ways that work and they may benefit from exploring some older solutions (see the box titled "Return of the Erie Canal"). For readers still uncomfortable with feudalism, other readings in this section and the next offer competing and, we like to think, complementary ideas that stress such key values as fairness, community resilience, psychological vibrancy, and social stability. In particular, consider Royce (chapter 13, this volume), who, from a century ago, discusses the benefits embedded in provincialism, a notion that has features in common with feudalism.

. . . Feudalism. The word usually generates an impression of brutish, servile peasants, hardly distinguishable from the soil they work, exploited by feudal lord and church alike. In contrast, we think of the Renaissance as sparkling with art, philosophy, and a newfound freedom—the age of light and discovery after the great darkness. Both impressions are half-truths. The history books were written by the winners of a century's long struggle between feudalism and the new spirit of commerce and science, which also enclosed the common fields and forced the peasants into the factories of the industrial revolution, the movement that earlier motivated the slave traders and fired the spirit of empire. Likewise, there is no doubt that there was exploitation in the feudal era, especially when it is remembered that feudalism was generally the pattern that came out of the collapse of the Roman Empire when the common people put themselves under the protection of a person powerful enough to defend them. (We might have feudalism in a similar form in the future if industrial society collapsed.) But to get a more correct picture of the era it is necessary to think of what this system evolved into: Chaucer, Boccacio, Dante, earthy peasants, and hard labor, it is true, but also humor, full-blooded sensuality, festivals, and a religious faith that could build the gothic cathedrals. A native aesthetic sense of peasants and craftsmen created the countryside, villages, and towns of the medieval era that fill the art and travel books fully as much as the grand monuments to the nation-states of later eras.

The feudal economy was based on religiously sanctioned contracts and mutual obligations rather than on the free market. It is true that the peasant had strict obligations to the feudal lord, to work in his fields and to provide other goods and services. But it was not a one-way street; the peasant was given the right to cultivate the feudal lord's land almost as if it were his own and the lord was also obligated to provide the peasant with defense, a mill, a church, and other services. These contractual obligations became enshrined as common rights and common laws that were protected in the folkmote, the village court of commoners which had authority over even the lord. Above all, this feudal system was stable, which was its greatest sin in the eyes of the reformers. It is true that this stability made it difficult for the feudal system to respond to the pressure generated by the reappearance of trade and a slowly increasing population. Commoners resisted any change in the common law for fear that they would lose out in the change, especially since most changes were proposed by lords with a mind to acquire the things that the nascent market offered. Now, however, the stability of the feudal system would again be an asset in the effort to establish a sustainable relationship with the environment.

It is hard to make a proposal that flies so directly into the face of a historical bias as strongly as does that against feudalism, but if this bias can be disregarded for a moment, the advantages of a new feudalism may become apparent. It resolves a number of problems that have so far plagued efforts to form effective alternatives to the ecologically unsound, socially threatening way of life that we know today. And a new feudalism would certainly not be a carbon copy of the old. Feudal contracts would be negotiated and entered into freely and would include safeguards. The new feudalism would benefit from the ecologically sound technology that is now available, and it could, to a certain degree, be adapted to our present attitudes and values.

If medieval Europe can be accepted as a general indication of the effect of the feudal system on the landscape, we could hope for a human and environmental richness that is in great contrast to rural areas in the United States, where isolated farms are scattered across vast, lonely landscapes, in which towns are often only a wide spot in the road, with a couple of stores, a gas station, and a church. The sense of abandonment, old age, and decay in rural America is oppressive, but at least it can be said that it offers an abundance of opportunity to anyone who can use the land more intensively.

Property Relationships

There is land, but many counter-culture efforts to go back to the land have floundered on the problem of obtaining land or the use of land. It is not too expensive to buy, but if it is purchased it uses up a good portion of the capital that is also needed to develop the community and to support it while it is getting underway. If one member of a group provides the land, he frequently becomes dissatisfied when other members of the group typically do not respond to his feeling that he should have some special role in the decisions of how the land is to be used; he begins to feel that he is being taken advantage of, that he is being used by the others. Even if the land is purchased jointly the problem occurs because more often than not some have contributed more funds than others.

Feudalism would avoid these problems. The landowner would receive deference from the community members in the form of the obligatory services to him, whether in farming, production, or services—whatever the initial contract called for. In return, the community members would receive the right to use and improve the land, to build dwellings and provide their basic needs without expenditures for land. Rights to the land would be based on individual family units, avoiding the conflicts that inevitably accompany communal ownership, enabling members to be industrious or easy-going as they please, without generating animosity from others, as long as obligations to the landlord and the community are fulfilled. The contract negotiated with the landlord would establish these relationships, and once established, they could not be broken without penalty. If a community member left, the improvements would revert to the landlord, a very important positive incentive to staying and providing continuity to the community. If the landlord wanted to end the experiment he would have to pay compensation to the community members, to purchase developments that he probably would not want.

Stability

It is fascinating to see how a feudal arrangement encourages stability, in contrast to the fluidity of a market system. In Europe the response was to carefully maintain and improve the land and to build homes and other structures that would provide well for the children and the children's children. Architectural styles evolved to high levels with great diversity from area to area in the absence of market incentives to reject traditional styles. How this contrasts with our new houses that barely last a single lifetime and where the current styles are much the same from one end of the country to the other, even though we theoretically have the potential to create great diversity.

Would there be individuals willing to become commoners in a new feudalism? There are certainly tremendous numbers of people who feel the pressure of present ways who would be interested in an alternative, but whether they would be interested in a new feudalism is hard to say. While many are unhappy with the pressure that is necessary to create our high standard of living and our constant stream of entertainments and diversions, I sense that only a few would be willing to give up the diversions that keep our minds occupied and the labor-saving devices that keep our muscles unoccupied. We would like to have our cake and eat it too, to have the good life without the high-pressure market economy that is necessary to produce it. It is going to take a certain degree of wisdom to see that any real alternative to the mainstream is going to be different but not necessarily better in every respect. We must make our choice.

Limits and Community

Unfortunately, few people in this country are aware of the values of a more static society, so pervasive has been the gospel of individualism and change for the last two hundred years. Characteristically, membership in a more static cultural unit greatly reduces the threats that are ever-present in our society: the pressures to perform on the job or in school, to be in style, to have friends, to be an "individual." Only a relatively static society can value the knowledge and experience of older people, and avoid the feeling of uselessness that afflicts the old in our dynamic society. A spirit of community, of common interests and expectations, is something about which we really know very little. The concept of the family should be the closest to it in our society, but even the family is under heavy stress. In a community, membership is all important. It provides mutual support and obligations. It discourages the loneliness of self-centered individualism, while permitting what is sometimes claimed to be a truer individualism. For with membership in the community assured there is no need to be concerned with conforming to superficial styles and mannerisms to maintain membership in a voluntary association. Static communities characteristically have the comic and the serious individuals, the loners and the garrulous, the foolish and the upright—the natural expression of personality in unthreatening circumstances, subject only to the limits established by the community.

It is the limits prescribed by the community which define it. It is sometimes said that the United States has no culture, which is only to say that we have very few of these defined limits. The result is that we are letting the situation find its own limits, many of which are turning out to be unsatisfactory. It would be preferable if limits could consciously be chosen by a community, and positive values created in support of them. Such limits are also what will make landowners feel confident enough to enter into a contract. Both of these objectives suggest the value of established and known religious systems.

Investment Benefits

Would landowners be willing to enter into such an arrangement? It would offer a landowner some important benefits he does not have now as a commercial farmer, which is a competitive and sometimes profitless business. Instead of being a single competitor in this scramble he would be to a degree free of it; he would no longer be alone in a vast environment but would have special status in the community that was developing on his land. If he had a particular interest in a certain form of political structure, philosophy, or religion, he could have the satisfaction of encouraging a community based on it. If a formal offer was made by a religious group in which he had confidence, even a rather conservative landowner might consider it.

Redefining Progress

. . . It is continuously said that man cannot go backwards. Perhaps this is because man defines "backwards" as opposite to the direction he is going at the time, and to go in such a direction would entail going against the internal logic of existing conditions and historic momentum. Man has often gone backwards, but only by force, because of invasion, internal collapse, economic reversal, or by natural forces that destroy the basis for a society's economic livelihood. For our society to redefine "progress" without such stimuli does seem to be impossible; it is true that there are very few examples of its ever having been done. For a small group of people, however, it might be possible, if they understood the nature of our society's problems and its values, to provide an alternative. An alternative community could be successful if it avoided the necessity of too radical a change in behavior and received firm institutional support of religious groups, but otherwise was as independent as possible.

Lasting Benefits

Such communities would be fundamentally different from the communes which are having such a difficult time achieving continuity and cohesion, let alone ecological wholeness. Yet there have been many such efforts, which suggests that if a less radical and less transitory alternative was available there would be more interest in it among a broader spectrum of the population. A new feudalism would offer many advantages: a slower, less frantic pace of life; work that is necessary for one's family and community rather than work that is psychologically unnerving and frequently superfluous; a community which, instead of being competitive and threatening, would provide a structure for existence on which one could count for help; a social environment that offers the possibility of being in control of one's life instead of being forced to cope with a world that is changing rapidly in undesired ways, and which every indication suggests will be harsher yet on our children. Finally, if we accept Hardin's statement, there will perhaps be the satisfaction of helping to establish the types of communities that are appropriate for the survival of mankind in the long run. I would like very much to suggest that religious bodies (and other groups with a vision of man) actively support the formation of new communities which embody the ethical systems on which they are founded.

9

Civic Agriculture*

Thomas A. Lyson

While localization may conjure up images of times long past or some distant future, much is indeed happening now. Hidden in plain sight, there is no better place to see localization in action than food.

Here sociologist Thomas A. Lyson illuminates the civic reinvention of local, agriculturally based enterprises. Lyson's analysis suggests that community-based food systems are reemerging throughout the country, largely without support from the centers of political and economic power. These new enterprises include community-supported agriculture in rural areas, farmers' markets in suburban communities, and urban agriculture in metropolitan areas. These findings are consistent with a central premise of this book, namely that localization in all sectors—food, energy, shelter, and transportation, for instance—will occur from within, as a result of the needs and desires of place-based communities to support themselves.

Notice, finally, how varied are the ownership and management structures Lyson identifies. The lesson for localization is that there is no one-size-fits-all prescription. Indeed, each response to declining net energy and environmental disruption is, and must be, locally tailored (see the box titled "Belo Horizonte").

The industrial type of agriculture produces most of America's food and fiber. However, a new form of civic agriculture that does not fit this conventional model of food production is emerging throughout the country and especially on the East and West Coasts. In this new civic

*Lyson, Thomas A. 2002. "Civic agriculture and community development." In *Civic Agriculture: Reconnecting Farm, Food, and Community*, © University Press of New England, Lebanon, NH. Excerpted and reprinted with permission, pages 84–98.

Belo Horizonte

In Belo Horizonte, Brazil, food security is now a citizen's basic right. Starting in 1993, the city secured this right through legislation and targeted programs to fix the local food system. At that time, malnutrition afflicted an estimated 20 percent of the city's children.

The local government's initiatives were multifaceted. For instance, despite the fact that the city set the food prices, farmers still did well. Andriana Aranha, head of the city's food initiatives, explains that by eliminating intermediate brokers and offering farmers prime locations throughout the city from which to sell their goods, the farmers profit through increased sales volume. In return, these farmers are expected to supply produce to the more impoverished citizens living in the city outskirts.

Through programs such as *Green Baskets*, the city pairs large food consumers (e.g., restaurants, schools) with local farmers. The focus of these programs is on locally supplied food to reduce transit costs. As an added benefit, the partnership provides patrons with nutrient-rich meals containing fresh produce and minimally processed foods.

Belo Horizonte also sustains the local food system by using community centers to distribute seeds and information about environmentally friendly farming techniques. For instance, the city has helped schoolhouses and neighborhoods incorporate open gardening spaces into their communities, which often double as outdoor classrooms.

The city is active but it doesn't directly administer programs, thus keeping food program costs low. The city works alongside private groups to create partnerships. In the end, Belo Horizonte exemplifies localization in practice, governmental and nongovernmental.

Source: Inspired by Frances Moore Lappé and Anna Lappé, *Hope's Edge: The Next Diet for a Small Planet* (New York: Tarcher/Putnam, 2003), 93–101.

agriculture, local agriculture is being reborn. This trend is most advanced and evident in the Northeast, especially New York, Vermont, and Massachusetts, where small-scale, locally oriented producers and processors have become keys in revitalizing rural areas of the region. These producers represent the vanguard of an important social trend.

To be sure, there is an emerging debate about whether civically organized local food systems can continue to expand and flourish in a globalizing environment. However, over the past 10 years, an accumulating body of research has begun to assess the benefits of small enterprises on the level of civic and community welfare. Communities that nurture local systems of agricultural production and food distribution as one part of a broader plan of economic development may gain greater control over their economic destinies, enhance the level of social capital among their residents, and contribute to rising levels of civic welfare and socioeconomic well-being.[1]

Civic agriculturalists and their enterprises are a varied lot, and no one set of characteristics perfectly defines these new producers. However, a profile that captures the tendencies of their operations compared to conventional agricultural producers can be constructed. . . . Civic agriculture is oriented toward local market outlets that serve local consumers rather than national or international mass markets. Farmers' markets, roadside stands, U-pick operations, and community-supported agriculture (CSA) are organizational manifestations of civic agriculture. Civic agriculture is seen as an integral part of rural communities, not merely as the production of commodities. The direct contact between civic farmers and consumers nurtures bonds of community. In civic agriculture, producers forge direct market links to consumers rather than indirect links through middlemen (wholesalers, brokers, processors, etc.). . . .

Farmers engaged in civic agriculture enterprises are concerned more with high quality and value-added products and less with quantity (yield) and least-cost production practices. Civic farmers cater to local tastes and meet the demand for varieties and products that are often unique to a particular region or locality. Civic agriculture at the farm level is often more labor intensive and land intensive and less capital intensive. Civic farm enterprises tend to be considerably smaller in scale and scope than those of industrial producers. Civic farming is a craft enterprise as opposed to an industrial enterprise. As such, it harks back to the way in which farming was organized in the early part of the 20th century. Civic agriculture takes up social, economic, and geographic spaces not filled (or passed over) by industrial agriculture.

Civic agriculture often relies on indigenous, site-specific knowledge and less on a uniform set of "best management practices." The industrial model of farming is characterized by homogenization and standardization of production techniques. The embedding of civic agriculture in the community and a concern with environmental conditions fosters a problem solving perspective that is site-specific and not amenable to a "one size fits all" mentality.[2]

Civic agriculture enterprises are visible in many forms on the local landscape. *Farmers' markets* provide immediate, low-cost, direct contact between local farmers and consumers and are an effective economic development strategy for communities seeking to establish stronger local food systems. *Community* and *school gardens* provide fresh produce to underserved populations, teach food production skills to people of all ages, and contribute to agricultural literacy. *Small-scale farmers*, especially organic producers, across the country have pioneered the development of local marketing systems and formed "production networks" that are akin to manufacturing industrial districts. *Community Supported Agriculture* (CSA) operations forge direct links between nonfarm households and their CSA farms. New *grower-controlled marketing cooperatives* are forming, especially in peri-urban areas, to more effectively tap emerging regional markets for locally produced food and agricultural products. *Agricultural districts* organized around particular commodities (such as wine) have served to stabilize farms and farmland in many areas of the country. *Community kitchens* provide the infrastructure and technical expertise necessary to launch new food-based enterprises. *Specialty producers and on-farm processors* of products for which there are not well-developed mass markets (deer, goat/sheep cheese, free-range chickens, organic dairy products, artisanal cheeses, etc.) and *small-scale, off-farm, local processors* add value in local communities and provide markets for farmers who cannot or choose not to produce bulk commodities for the mass market. What these civic agriculture efforts have in common is that they have the potential to nurture local economic development, maintain diversity and quality in products, and provide forums where producers and consumers can come together to solidify bonds of community.

Community-Supported Agriculture

Community-supported agriculture consists of a group of individuals or families who commit resources (money and/or labor) to a farmer and become, in essence, shareholders of the farm.[3] In return for their investment, the shareholders receive part of what the farm produces that season. CSA shareholders provide farmers with the money they need to finance their operations before the growing season begins. In this way, the shareholders incur along with the farmers both the risks and the benefits of food production.

Most CSAs offer their members (shareholders) a variety of fruits, vegetables, herbs, and flowers in season. Some CSAs also produce eggs, milk, meat, baked and canned goods, and even firewood. In a recent study of how CSAs work, Bruno Dyck found that most ranged in size from 35 to 200 shareholders. A typical box of food, usually distributed weekly, one per shareholder, held between five and ten pounds of fruits and vegetables. Consumers pay from $10 to $35 per week, and an average share for a season was $346.[4] . . .

Each CSA is organized to meet the needs of its shareholders. CSAs vary according to the level of financial and labor commitments of their members, their decision-making structures, ownership arrangements, and methods of payment and food distribution. Four of the most common forms of CSA include:[5]

1. Farmer-directed CSAs: The producer organizes the CSA and takes major responsibility for managing it. Shareholders are seen as "subscribers" and have minimal involvement in the day-to-day operation of the farm. For a cash share, paid before the season begins, subscribers to the farm receive a box of food and other agricultural products on a weekly basis throughout the growing season.
2. Consumer-directed CSAs: A group (community) of consumers organize the CSA and then recruit a farmer to produce for them. Decisions regarding what will be grown and under what methods are made by the consumers, though the farmer is typically brought into these discussions. Labor is sometimes provided by CSA members.
3. Farmer-coordinated CSAs: Two or more producers pool their resources and expertise to produce a wide variety of food and agricultural products for an expanded group of consumers. Farmer-coordinated CSAs might include milk, eggs, and meat in addition to the fruits and vegetables typically associated with CSA operations. Each producer in the coordinated CSA specializes in one product or a set of products. A network of CSA producers thus meets a wider set of needs than any one farmer could individually.
4. Farmer-consumer cooperatives: Producers and consumers join together to purchase land and equipment for the CSA. Decisions regarding what to grow and under what conditions are made jointly by the farmers and the shareholders.

In all four CSA types, the farmer develops a production plan and budget. This is sometimes done with shareholder input and sometimes without it. The budget covers all the anticipated costs of production, including a fair wage for the farmer and other employees of the CSA. The total costs are then divided among the number of shares to be sold. Some CSAs have developed arrangements to subsidize low-income shareholders and to divert part of the food produced to food banks and other emergency food outlets.

Although CSAs take many forms, all are committed to establishing and maintaining a more local and just food system. CSAs combine a concern over land stewardship with an imperative to maintain productive and profitable small farms. In a 1995 study of CSA shareholders, Cynthia Abbott Cone and Ann Kakaliouras found that environmental and community concerns were more important than the price of food as reasons why the shareholders joined a CSA.[6]

CSAs are an important part of civic agriculture. They strengthen the local food economy and preserve farmland. A web of connected and cooperatively organized CSAs could represent a real and viable alternative to the mass-produced, homogeneous, imported produce found in most supermarkets today. . . .

Restaurant Agriculture

Restaurant agriculture, which is sometimes referred to as "culinary agriculture," involves a production and marketing strategy that brings together farmers and chefs in a mutually supporting and beneficial relationship. Farmers and chefs work together to develop cuisines that draw on the unique aspects of local agriculture. Over the course of a year, chefs who participate in restaurant-supported agriculture often change their menus to incorporate different products as they become available during the growing season.

Farmers who commit to restaurant-supported agriculture provide restaurant owners and chefs high-quality fruits, vegetables, meats, and dairy products. For these consumers, low price is not a major concern.[7] Building trusting relationships between restaurant owners and farmers is the key to success. Maintaining trust requires developing a high degree of familiarity among farmers, chefs, and restaurant owners. Annual visits to one another's businesses are but one way that farmers and chefs come to understand each partner's needs.

Farmers' Markets

Before large supermarkets became a fixture on America's food landscape, farmers' markets provided consumers with a wide array of fresh, local produce. Most cities and towns had at least one farmers' market. In large cities, local farmers' markets catered to the particular tastes and wants of residents who lived in ethnic neighborhoods.

The number of farmers' markets in the United States began to decline in the 1920s with the advent of the modern supermarket. By the 1970s, the number of farmers' markets reached its nadir, with fewer than one hundred still operating. Although the numbers of farmers' markets plunged, especially after World War II, they never totally disappeared. Like an old idea whose time had come again, farmers' markets began what has become a rather remarkable rebound.[8] The most recent figures from the U.S. Department of Agriculture show over three thousand farmers' markets in 2002. [USDA: 7,175 in 2011] In New York City alone, farmers' markets operate in twenty-eight different locations.

Farmers' markets offer a convenient outlet for producers who cannot or will not develop linkages to the mass market. They also fill an important niche for consumers who value quality, freshness, and variety over quantity and uniformity in the food they purchase. Most farmers' market vendors pride themselves on selling varieties of fruits and vegetables that cannot be found in the typical mass-market grocery store.

While health- and food-conscious consumers are purchasing more fresh fruits and vegetables than ever from farmers' markets, local government officials see farmers' markets as engines of economic and community development. In a survey of 115 farmers' market vendors, we found that farmers' markets provide a venue for [two] groups of producers. First, for some traditional full-time farmers, farmers' markets can provide a steady source of income. Over the past several decades, the number of marketing alternatives in the processing sector has diminished considerably. The number of small-scale food processors, those most likely to articulate with small farmers, has decreased almost everywhere. For small-scale producers, farmers' markets may represent an economic lifeline.[9]

Part-time growers and market gardeners represent another segment of the agricultural community who benefit from farmers' markets. Farmers' markets allow these producers to sell their products directly to the consumer and can supplement other marketing outlets such as roadside stands, direct-mail marketing, and U-pick operations. . . .

As social institutions and social organizations, farmers' markets can be important components of civic agriculture. As bridges between the formal and informal sectors of the economy, they enable individual entrepreneurs and their families to contribute to the economic life of their communities by providing goods and services that may not be readily available through formal market channels. They embody what is unique and special about local communities and help to differentiate one community from another. . . .

Urban Agriculture, City Farming, Community Gardens

To many people, urban agriculture and city farming sound like oxymorons. Most of our food and agricultural products are produced on farms that are located far from the bright lights of the city. Over the past twenty years, however, farming opportunities have been sprouting in the nation's metropolitan areas.

According to the Canadian bioecologist William Rees, urban agriculture includes growing crops and raising some forms of livestock in or very near cities for local consumption. Community or urban gardens are probably the most visible form of urban agriculture. The first community gardens were organized by the mayor of Detroit in the 1890s to help families cope with the effects of the economic depression of that era. Throughout history, whenever there has been a shortage of food or money, community gardens have flourished. The Liberty and Victory Gardens of World Wars I and II served to supplement the food rationing imposed on the nation at that time. Community gardens also surfaced during the Great Depression as part of the nation's "emergency food system."[10]

Even today, many community gardens are critical sources of food for low-income people. While it costs next to nothing to garden, the average urban garden produces about 540 pounds of food a year. If purchased in a grocery store, the fruits and vegetables grown in the average garden would cost almost $500.[11] But many community gardens today are more than just sources of almost-free food for poor and low-income people. Many observers have noted that community gardens are a "way for people to work together, socialize and talk with their neighbors. Users plan, construct, and maintain the space, thus building community relations at the same time they save money and lower their cost of living."[12] Urban gardens nurture not only plants and animals but people and their cultures as well.

Urban gardens can teach entrepreneurial skills and spawn and sustain a broad range of new employment opportunities. Not only do community gardens teach horticultural skills, but in some cases they encourage new marketing initiatives, environmental management activities, and community development processes.

Most importantly, William Rees notes that "urban farming can contribute to the rebirth of civil society and development of community as neighbors cooperate in the establishment, management and supervision of community-owned or accessible garden plots."[13] Recently, the *Trends Journal* noted that urban gardening will likely be among the most visible manifestations of a new urban revival—a revival that sees neighborhoods and the groups and organizations embedded in them as the building blocks of a vital civic community.[14]

Measuring Civic Agriculture

Many civic agriculture enterprises exist off the radar screens of most federal and state agencies. Only recently, for example, has the USDA begun collecting and distributing information on farmers' markets.[15] And in 2002, the USDA began an organic certification program. However, it is unclear how many "organic" farmers will actually participate in the National Organic Program, in part because many smaller-scale producers already have a customer base for whom national certification is not needed.[16]

Despite the lack of reliable national statistics on civic agriculture, various organizations around the country have been trying to monitor the growth of civic agriculture. The Community, Food, and Agriculture Program at Cornell University, for example, regularly updates its listing of civic agriculture and food enterprises in New York.[17] . . . [In New York state] the number of farmers' markets grew from 6 in 1964 to 269 by 2002. The number of small-scale organic farmers increased over tenfold between 1988 and 2002, while the number of community gardens increased threefold since 1978. The number of farmers selling directly to the public increased by nearly 600 between 1992 and 1997. Today nearly one in seven farmers in New York sells directly to the public.

Civic agriculture activities such as urban gardens, farmers' markets, roadside stands, and CSAs, as aspects of the civic community, become a powerful template around which to build non- or extramarket relationships between persons, social groups, and institutions that have been distanced from each other. Indeed a growing number of community groups across the United States are recognizing that creative new forms of community development, built around the regeneration of local food systems, may eventually generate sufficient economic and political power to mute the more socially and environmentally destructive manifestations of the global marketplace. A turn toward a more civic agriculture is both theoretically and practically possible. Indeed, the seeds have been sown and are taking root throughout the United States. Civic agriculture represents a promising economic alternative that can nurture community businesses, save farms, and preserve farmland by providing consumers with fresh, locally produced agricultural and food products.

Notes

1. Green, Joanna and Duncan Hilchey. 2002. *Growing Home: A Guide to Reconnecting Agriculture, Food and Communities.* Ithaca: Community, Food and Agriculture Program, Department of Rural Sociology, Cornell University. See also Lyson, Thomas A. and Judy Green. 1999. The Agricultural marketscape: A framework for sustaining agriculture and communities in the Northeast. *Journal of Sustainable Agriculture* 15: 133–150.

2. Civic agriculture should be viewed as an ideal type. Ideal types are mental constructs against which empirical cases can be compared. The characteristics of the ideal type of civic agriculture were drawn from a number of sources including Wilkins, Jennifer. 1995. Seasonal and local diets: Consumers' role in achieving a sustainable food system. *Research in Rural Sociology and Development* 6:149–166; Center for Rural Affairs. 1988. Agriculture: A foundation for rural economic development. *Center for Rural Affairs Newsletter*, November SR-5; Waters, Alice. 1990. The farm–restaurant connection. In *Our Sustainable Table.* ed. Robert Clark, 113–124. San Francisco, CA: North Point Press; Bird, Elizabeth Ann, Gordon L. Bultena, and John C. Gardner, eds. 1995. *Planting the Future: Developing an Agriculture that Sustains Land and Community.* Ames, IA: Iowa State University Press; Kloppenburg, Jack J. 1991. Social theory and the de/reconstruction of agricultural science: Local knowledge for an alternative agriculture. *Rural Sociology* 56: 519–548; Kneen, Brewster. 1993. *From Land to Mouth: Understanding the Food System.* Toronto, CN: NC Press Ltd; Johnston, Thomas R. and Christopher R. Bryant. 1987. Agricultural adaptation: The Prospects for sustaining agriculture near cities. In *Sustaining Agriculture Near Cities.* ed. William Lockeretz. Ankeny, IA: Soil and Water Conservation Society; and Lyson, Thomas A., Gilbert Gillespie, and Duncan Hilchey. 1995. Farmers' markets and the local community: Bridging the formal and informal economy. *American Journal of Alternative Agriculture* 10:108–113.

3. For background on community-supported agriculture see Fieldhouse, Paul. 1996. Community shared agriculture. *Agriculture and Human Values* 3:43–47; and Cone, Cynthia A., and Andrea Myhre. 2000. Community-supported agriculture: A sustainable alternative to industrial agriculture? *Human Organization* 59:187–197; and Kittredge, Jack. 1996. Community-supported agriculture. In *Rooted in the Land: Essays on Community and Place.* eds. William Vitek and Wes Jackson. New Haven, CT: Yale University Press.

4. Dyck, Bruno. 1992. Inside the food system: How do community supported farms work? *Marketing Digest* (August): 2.

5. Appropriate Technology Transfer for Rural Areas (ATTRA) is an excellent source of radical information on CSAs. See, for example, Greer, Lane. 1999.

"Community supported agriculture." *Business Management Series*. Fayetteville: Appropriate Technology Transfer for Rural Areas (ATTRA).

6. Cone, Cynthia A., and Ann Kakaliouras. 1995. "The quest for purity, stewardship of the land and nostalgia for sociability: Resocializing commodities through community supported agriculture." *CSA Farm Network*. Stillwater, NY: CSA Farm Network.

7. Restaurant-supported agriculture (RSA) is a relatively recent phenomenon. Only a few studies have begun to examine the potential economic and social impacts of RSA. The following are useful places to learn more about RSA. ATTRA, www.attra.org/attra-pub/altmeat.html; Berkshire Grown, www.berkshiregrown.com; and Chef's Collaborative, www.chefnet.com/cc2000. See also Green and Hilchey 2002.

8. For some background information on farmers' markets see Lyson, Gillespie, and Hilchey 1995. A good empirical study of farmers' markets is Hughes, Megan E., and Richard H. Mattson. 1992. "Farmers' markets in Kansas: A profile of vendors and market organization." *Report of Progress 658*. Manhattan, KS: Experiment Station, Kansas State University.

9. See references in note 8 above. See also Hilchey, Duncan, Thomas A. Lyson, and Gilbert W. Gillespie. 1995. "Farmers' markets and rural economic development." *Community agriculture development series*. Ithaca, NY: Cornell University, Farming Alternatives Program, Department of Rural Sociology.

10. Rees, William. 1997. "Why urban agriculture?" *City Farmer*. Urban agriculture notes and Woodsworth, Alexandra. 1995. *Community gardening: A Vancouver perspective*. Vancouver, CN: City Farmer Canada's Office of Urban Agriculture.

11. Woodsworth, 1995.

12. The American Community Garden Association is an important source of information about urban gardens, community gardens, and school gardens: see http://www.communitygarden.org. The quote is from "Comprehensive Plans, Zoning Regulations, and Goals Concerning Community Gardens and Open Green Space from the Cities of Seattle, Berkeley, Boston, and Chicago" on the American Community Garden Web site.

13. Rees, 1997.

14. *Trends* Journal 1997 as cited in *Yes!* 1997. Top Trends '97. *Yes! A Journal of Positive Futures*. (Spring).

15. Information on farmers' markets can be found at http://www.ams.usda.gov/farmersmarkets/.

16. The National Organic Program homepage is at http://www.ams.usda.gov/nop/indexIE.htm.

17. The Community, Food, and Agriculture Program (CFAP) was founded in 1985 as the Farming Alternatives Program. CFAP works with agriculture and food producers and community partners to promote food and agriculture systems that sustain and strengthen farm families, local communities, and natural resources. See http://www.cfap.org.

10

A Whole New Way of Life: Ecovillages and the Revitalization of Deep Community*

Karen Litfin

Perhaps the most ambitious and self-conscious localizers are those who are creating ecovillages. Around the world, North and South, among rich and poor, diverse experiments are being conducted, each driven in part, it seems, by an expectation that life will be changing fundamentally and it's time to get ready.

Political scientist Karen Litfin has studied fourteen such communities on five continents, acquiring firsthand a sense of the practices and philosophies of those doing localization. Note that while she finds that these efforts fit no single profile ideologically or culturally, she finds them unified in their commitment to an affirmative politics, rather than a politics of protest. Perhaps most significantly, and maybe most surprisingly, she concludes that many ecovillages are best understood as "dynamic nodes of global engagement," not isolated enclaves of extreme localism. Finally note that, like so many current localizing efforts, "ecovillages have done [all this] so quietly that few people are even aware of their existence."

Climate change, the mass extinction of species, resource depletion—all of these point to one colossal conclusion: our everyday actions are unraveling our planet's life-support systems. Yet the greatest success stories of top-down politics are usually too little, too late. Despite a plethora of treaties, institutions, and organizations, our home planet's life-support systems are unraveling at an alarming rate. Yes, we need new laws and new products. But these piecemeal solutions only tinker at the margins when what we need are whole new ways of being that integrate the social, economic, ecological, and personal dimensions of life. . . . I asked

*This is the first publication of this article. Permission to reprint must be obtained from the publisher.

myself, "Who is devising the far-reaching solutions to the sustainability crisis, and applying them at the level of lived experience?" This question led me on a nine-month sojourn around the world to study ecovillages. I visited fourteen communities on five continents, interviewing ten residents in each in order to understand what works and what doesn't.

In response to the looming megacrisis, ecovillages have quietly taken root all over the world, in rich and poor countries alike. Their populations range from 20 to 2,000. Their beliefs are rooted in every major world religion, plus paganism and atheism. These communities are not just isolated enclaves; they are intensely engaged in public education and mutual learning. Since 1995, with the formation of the Global Ecovillage Network (GEN) (see Global Ecovillage Network 2004), hundreds of ecovillages around the world have come together in order to share and disseminate their knowledge. Network members include large networks like Sarvodaya (11,000 villages applying ecological design principles in Sri Lanka) and the Colufifa network of 350 villages in Senegal; the Ladakh project on the Tibetan plateau; ecotowns like Auroville in South India and the Federation of Damanhur in Italy; small rural ecovillages like Earthaven in North Carolina and Huehuecoyotl, Mexico; urban rejuvenation projects like Los Angeles Ecovillage and Christiania in Copenhagen; permaculture design sites such as Crystal Waters, Australia, and Barus, Brazil; and educational centers like the Centre for Alternative Technology in Wales. Dawson (2004) notes that these communities trace their roots to

- The ideals of self-sufficiency and spiritual inquiry embodied in monasteries, ashrams, and the Gandhian movement
- The "back-to-the-land" movement and, later, the cohousing movement
- The environmental, peace, and feminist movements
- The participatory development movement
- The alternative education movement

Ecovillages have taken root in tropical, temperate, and desert regions. Their specific practices vary according to cultural and ecological context, but the common thread is a shared commitment to personal, social, and ecological harmony. Beneath this commitment to social and ecological sustainability, one may discern a worldview premised on holism and radical interdependence—a radically different perspective from the assumptions underlying modern consumerism.

Unlike the larger environmental movement, ecovillages are not primarily concerned with protesting against state and industry lassitude. Rather than waiting for the revolution, they are *prefiguring* a viable future by creating parallel structures for self-governance in the midst of the prevailing social order. Their resource consumption is far lower than their home countries' averages. And because they have found creative ways of limiting their participation in the global economy, their average income is low—dispelling the myth that sustainability is a luxury for the rich. Yet theirs is a culture of abundance, not deprivation. Focusing on the most practical and down-to-earth issues of life—food, energy, transport, housing, and, above all, community—this movement embodies a kind of hands-on, do-it-yourself politics, a politics of "yes."

Ecovillages offer one response to the call for a new kind of environmentalism. An increasingly vociferous chorus of scholars, activists, and policy analysts warns that the prevailing economic and political system—including traditional activism—has broken down. John Dryzek, a preeminent scholar in the field, concludes that capitalism, interest-group politics, and the administrative state are "thoroughly inept when it comes to ecology" (2003, 474). Yale's Gus Speth (2008, 86) concludes in his *Bridge at the Edge of the World* that "working only within the system will, in the end, not succeed when what is needed is transformative change in the system itself." In their influential manifesto, *Death of Environmentalism*, Michael Shellenberger and Ted Nordhaus (2004) call for a radical, holistic, and affirmative rethinking of sustainability. Ecovillages have done just this, but they've done it so quietly that few people are even aware of their existence.

People long for a sense of integrity and deep connection, a longing that is all the more acute amid our harried, technologically mediated modes of living. After living in fourteen ecovillages on five continents, I have come to believe that this elusive sense of belonging is best nurtured when interdependence is no longer just an idea but is grounded in everyday life. We might install compact fluorescent bulbs and buy organic, but our actions are more palliative than efficacious. They are too small, both in material terms and as reflections of our humanity. This is because the creeping global crisis is not only a physical or political problem "out there." It is a crisis of meaning "in here." Therefore the truly effective alternatives will not only work pragmatically, they will also nurture within us a sense of deep belonging—to the planet and to each other. Given the scale and scope of the challenges ahead, our hunger for integrity can only be satisfied by the integration of ecology, economics, human solidarity, and inner meaning in our daily lives. This is precisely what ecovillages aim to do.

This chapter offers an overview of the holistic approach to living that I experienced during my travels to fourteen very diverse ecovillages. The printed page, however, is not the optimal medium for conveying the colorful dynamism of these places. For those who wish to have a visual sense of ecovillages, I have posted a short video, "Seed Communities: Ecovillage Experiments Around the World," at http://www.youtube.com/watch?v=MtNjZaXDGqM.

In the following pages, I first explore the holistic understanding of life that underpins the ecovillage vision, with a special focus on its roots in permaculture. The next section investigates how the individual and the global are constituted within the holistic ontology that informs the ecovillage movement. I then discuss the intersection of lifestyle politics and transnational activism. Finally, I conclude with a brief inquiry into the question of effectiveness.

A Holistic Approach to Human Systems

Central to ecovillage life is the creation of virtuous cycles, as opposed to vicious cycles, which regenerate the land, enliven the community, and sustain its members in a cohesive whole consisting of integrated human and natural systems. Graywater and kitchen waste are recycled into community gardens; human manure is composted into landscape soil; rainwater is harvested for garden and home use; woody waste from community forests warms the homes of the residents. To the extent that ecovillages are able to generate a local economy based on community resources, money circulates internally and automobile use decreases. One expression of this holism, one that predates and powerfully informs the ecovillage movement, is permaculture.

Developed in the 1970s in Australia by Bill Mollison and David Holmgren, permaculture focused on small-scale sustainable agriculture and integrated systems at the household level (Holmgren and Mollison 1978). Over the next two decades, the concept was adapted to diverse social and ecological contexts around the world. Its original land-use focus was extended into every aspect of *sustainable culture*, including matters of social and spiritual well-being. Most important, permaculture promotes bottom-up social change rooted in design principles observable in nature. It locates the drivers for social, economic, and cultural change at the individual and the household level. Ecovillages have expanded this vision from the household to the community level.

Permaculture has been tailored to local contexts by ecovillages around the world and its core concepts infuse the ecovillage movement:

1. Design from nature,
2. Catch and store energy,
3. Make the smallest intervention necessary,
4. Use small and slow solutions,
5. Apply self-regulation and accept feedback,
6. Produce no waste,
7. Use and value diversity,
8. Integrate rather than segregate. (Holmgren 2002, viii)

Drawing on the science of ecology and an acute awareness of industrial society's fossil fuel dependency, permaculture emphasizes the wise husbandry of energy resources. Living systems, including human ones, are understood in energetic terms: food, trees, soil, buildings, modes of transportation, water—*all* are embodiments and conductors of energy. The consequences ramify across systems. In food production, soil fertility is enhanced through composting, thereby minimizing the loss of energy. Likewise, the perennial edible landscapes associated with permaculture, including food trees, berries, and herbs, decrease the energy input required for food production. Permaculturalists are attentive to the health of forests because trees, more than any land species, accumulate biomass rapidly and thus represent a tremendous storehouse of energy. Not surprisingly, most of GEN's Living and Learning Centers offer permaculture courses.

Nature knows no waste, yet this human construct is a foundational premise of modern industrial societies. Permaculture seeks to close the cycles of production and consumption, thereby minimizing waste, through such low-energy technologies as the composting of agricultural and human waste, the use of renewable energy, the recycling of gray water into food production, and the use of natural building materials such as cob, straw bales, and compressed earth blocks. Ecovillages, which grew in part from the permaculture movement, are demonstration sites for these and other low-energy, earth-friendly technologies.

Permaculturalists foresee the eventual elimination of the "fossil fuel subsidy" and a "descent culture" moving toward a low-energy, sustainable future in the coming centuries (Holmgren 2002, xxix). I found variants of this macrolevel perspective about energy descent in most of the ecovillages I visited. Their use of low-energy building materials, their commitment to wind, solar, and biomass energy systems, and their frequent preference for using human (as opposed to machine) energy—all of these practical applications come as a natural corollary to this macrovision. Most ecovillagers I spoke with were acutely aware of world energy trends. So I was not surprised to find some apocalyptic and neoprimitivist thinking among them, but I was surprised to find relatively little of it. Some of the more optimistic prognoses stemmed from a belief that our tightly networked, information-rich culture will facilitate the rapid innovation and spread of sustainable living practices. David Holmgren (2002, 22), one of permaculture's founders, articulates this view of modern society as a "fast-breeder system that generates new information, knowledge, innovation and culture."

GEN got started just as the Internet was gaining widespread use. Given the global nature of the movement and its aversion to fossil fuel usage, the Internet has been key to disseminating information, sharing best practices, and organizing regional and global conferences. Ecovillages are therefore hubs of learning. A constant flow of exchange with other communities and visitors moves across their porous boundaries. I was intrigued by some of the resulting combinations of low-tech and high-tech: computer programmers, web designers, and telecommuters living in solar-powered treehouses, adobe cabins, and rammed-earth huts. Ecovillages are not isolated enclaves of escapists; rather, they are dynamic nodes of the information society.

Despite the *eco-* in *ecovillage*, most people told me that their greatest joys and challenges come from the social aspects of community life. Besides being laboratories for ecological living, ecovillages are also experiments in radical democracy. Most of them operate by consensus decision making. This doesn't mean that everybody necessarily agrees on everything; it only means that people must be sufficiently satisfied with decisions not to block them. The basic idea is that the best decisions will be made when minority views, rather than being overruled by the majority, are considered and incorporated into better proposals. When it works well, consensus decision making is an inclusive, cooperative, and egalitarian process. When it doesn't: welcome to the flipside of mainstream democracy—tyranny of the minority.

While human relationships are the most fractious aspect of community living, they are also the most potentially rewarding. Fortunately, today's communities are the beneficiaries not only of the information society but also of a host of skills in consensus decision making, meeting facilitation, and nonviolent communication. Consequently, the trends are towards stronger interpersonal clarity, shorter meetings, greater transparency and trust, and more individual freedom. These trends are amplified by the information-rich context that ecovillages inhabit. Permaculture offers design principles for social as well as ecological sustainability: apply self-regulation and accept feedback; use and value diversity; integrate rather than segregate.

The work of connectivity entails recognizing and appreciating how the world looks from the vantage points of others: other people, near and far, and nonhuman others. The primary instrument for this work is empathy—what the Dalai Lama calls "the master emotion"—and the work is simultaneously political, social, economic, ecological, and spiritual.

Holism: Linking Self to World

The holism of ecovillages is not only pragmatic in its design from nature, but also serves as a source of meaning for the individual. Rather than becoming lost in the whole, which from the atomistic perspective of modernity would be the inevitable fear, each individual inhabits the center of a series of concentric circles beginning with home and extending to community, ecosystem, nation, and planet. This holism challenges the possessive individualism of modern consumer culture by integrating person and planet within the context of community. Recognizing their own complicity in replicating the social structures that threaten to unravel Earth's life-support systems, ecovillagers accept responsibility for their own lives even as they seek to invent alternative social structures.

Yet some critics of "lifestyle strategies" view ecovillages as self-indulgent, escapist, and ineffective responses to the powerful global structures that perpetuate socioeconomic injustice and environmental degradation (Fotopoulos 2000). This perspective fails to appreciate the radical consequences of a holistic worldview. From a mechanistic perspective, neither the lone individual nor tiny communities have the power to effectively counter enormous institutions; they are simply too small. From a holistic perspective, however, the networks of interdependence within a system are so intricate and tightly interwoven that one can never say for certain that an individual's actions will be insignificant. In the words of ecovillage proponent Ross Jackson (2000 37–38), "We are like individual molecules being perturbed on a global scale by a technology that is too powerful for us to handle in the dream state. We will have to wake up soon." Jackson believes that the vehicle for this awakening will be "self-organizing systems: a grassroots local initiative, a decentralized explosion of energy with a global vision." From this perspective, it is quite possible that effective responses to the sustainability crisis could come from individuals and communities—especially if these local entities are globally linked.

The global ecovillage movement is politically unconventional for another reason: it is an affirmative movement, not a protest movement. Rather than resisting what they oppose, ecovillages are building an alternative from the ground up, an approach that makes sense from a holistic perspective. If the world is radically interdependent, then establishing alternative social structures with the potential to ramify may be more effective than simply saying no to unsustainable systems. Moreover, there is the question of integrity. Ecovillagers tend to be wary of the scale and complexity of global social and technological systems, believing that responsible action is most viable at the scale of the individual and community.

Unlike earlier intentional communities and back-to-the-land experiments, ecovillages are not isolated enclaves. They tend to be active in local, national, and transnational politics. Political engagement follows from a holistic understanding of the world because individuals and communities are always embedded in larger ecological and human systems. Most of the ecovillagers I encountered saw themselves as engaged participants in planetary socioecological systems rather than as utopian fugitives. On a principled level, they view their lives as pragmatic responses to the interrelated global dynamics of North-South inequity, global commodity chains, structural violence, and fossil fuel consumption. At the level of action, ecovillage activists have been prominent players in the movements for peace, human rights, and global justice.

Ecovillages are a conscious response to socioecological realities in the global North and South. In affluent countries, ecovillages seek to reinvigorate social life and decrease material consumption. In developing countries, ecovillages aim to preserve village life and enhance material living standards in a sustainable manner, thereby providing an alternative to poverty, urbanization, and corporate-led globalization. Because a large number of people in developing countries still live in village settings, the foremost concern of Third World ecovillage movements is to incorporate appropriate technologies and community-building skills into existing villages. In industrialized countries, where small-scale communities have been overtaken by the forces of urbanization and suburbanization, ecovillagers are establishing islands of relative sustainable community in a sea of affluence, alienation, and wastefulness. Yet, like their counterparts in the South, they seek to meld the best of new appropriate technologies (e.g., solar, wind, and biofuel energy sources) with traditional practices (e.g., mud building and organic farming). In both contexts, the village model is seen as a compelling response to the global economy, succumbing to neither the affluence of the overconsuming North nor the grueling poverty of much of the South. Thus, the ecovillage movement represents an effort to forge a third way, applying and integrating the best practices of North and South with sensitivity to local context.

At the level of political action, opposition to economic globalization serves as a rallying point for ecovillagers around the world. Yet even in their opposition, they move beyond the politics of protest. For instance, during the 2005 G8 summit in Scotland, Findhorn members helped to create a temporary ecovillage as a counterpoint to the unsustainable policies being promoted at the summit. The demonstration ecovillage included composting toilets, graywater systems, solar panels and wind turbines, and kitchens serving fair trade, local, and organic food. The G8 action "illustrates the primary gift of ecovillages to the wider sustainability family; namely, the impulse to move beyond protest and to create models of more sane, just and sustainable ways of living" (Dawson 2006, 38).

Ecovillages are heavily involved in international peace, humanitarian relief, and solidarity work. Sarvodaya, GEN's largest member, has been an active peace broker in Sri Lanka's long-term civil war. Both Sarvodaya and Auroville assisted in tsunami relief work. Plenty International, an NGO created by The Farm in Tennessee, specializes in bringing soybean agricultural training to the Third World. Tamera ecovillage in Portugal, which considers itself a "research center for lived peace," is involved in conflict resolution work in Colombia and the Middle East. Ecovillagers in Denmark have partnered with ecovillagers in Senegal to send thousands of bicycles there. Far from being exclusive enclaves of escapists, many ecovillages are dynamic nodes of global engagement.

This engagement is also evident in the many ecovillage educational centers—especially GEN's five regional Living and Learning Centers. These full-immersion, hands-on learning centers teach a compendium of sustainable living skills, including organic gardening, living systems for water cycling, renewable energy sources, earth-friendly building from local materials, alternative economics, and community building and conflict resolution skills. Local skills and knowledge are woven together with permaculture methodology, forging workable solutions to the specific problems faced in each culture (Snyder 2006).

The holistic vision of ecovillages, while focused on sustainable living practices at the individual and community level, is also a globally engaged vision. At all levels, from the individual to the global, the focus is on establishing ecologically and socially viable alternatives on the ground.

The Power of the Small

Even if ecovillages are internationally engaged and demonstrating the practical viability of a holistic worldview, their numbers and influence are so small that one must wonder if they should be taken seriously. They might provide a cozy green life for a few lucky individuals, but can they ever hope to be effective as a transformative force in the face of global capitalism and the culture of consumption? If anything, the problems of environmental degradation, social alienation, and North-South inequity have only worsened since the inception of the ecovillage movement. Yet, on a smaller spatial scale and a longer temporal scale, there are good reasons to be impressed by the actual and potential effectiveness of ecovillages.

First, these communities demonstrate concretely that material throughput can be substantially reduced while enhancing the quality of life. Many of the ecovillagers I interviewed live on a tiny fraction of the average income for their countries. Residents of Earthaven, an ecovillage in rural North Carolina, live comfortably on less than $8,000 per year. Likewise, members of Sieben Linden in Germany live well on less than €10,000 per year. These numbers undercut two widespread assumptions: that high levels of consumption are correlated with well-being and that sustainability is a luxury for the affluent. Second, ecovillages show that we can live well while dramatically reducing our consumption and waste. Residents of Crystal Waters Permaculture Village in Australia, for instance, have been able to reduce their per capita solid waste by 80 percent over the regional average. A 2003 study by the University of Kassel found that the carbon footprint of Sieben Linden was 72 percent below the German average. Even in the suburban Ecovillage at Ithaca, a very middle-class U.S. ecovillage, the per capita ecological footprint is 40 percent lower than the national average.

While ecovillages may show that another world is possible on a minuscule scale, the question remains: Can global systemic change come about through a network of communities committed to social and ecological sustainability? The short answer is: We don't know. In the absence of more far-reaching forms of political engagement directed toward structural change, the strategy of lifestyle politics is a doubtful one. Yet, as we have seen, many ecovillagers are also working for structural change.

Ecovillages are not *the* answer to the sustainability crisis. They are just *one* answer—and we need all the answers we can get. Ecovillages are seeds of hope broadcast across the global landscape, small and sparsely sown. Time is short. We can't all go out and build new ecovillages, nor should we. We can, however, apply the lessons of the ecovillage in our homes, neighborhoods, schools, and workplaces. In essence, this is what the Transition Towns Movement is about: a scaled-up and highly dispersed rendition of the ecovillage model. The basic principle is simple: sharing—sharing material resources, ideas, dreams, skills, stories, joys, and sorrows. We don't need to live in ecovillages to abide by this principle. And if we did live by this principle, our communities would look more like ecovillages.

If existing ways of living are not sustainable, they will cease. The only questions are when and how. Whether the demise of the current order is precipitous or gradual, any successful experiments will become enormously salient. From that standpoint, even if the seeds of the ecovillage movement seem sparsely sown and its successes modest, the whole new way of life ushered in by this movement takes on a new light.

References

Dawson, Jonathan. 2004. Wholesome living. *Resurgence* 225 (July/August): 32–38.

Dawson, Jonathan. 2006. *Ecovillages: New Frontiers for Sustainability*. Devon, UK: Green Books.

Dryzek, John. 2003. "Ecology and Discursive Democracy: Beyond Liberal Capitalism and the Administrative State." In David Pepper, ed. *Environmentalism: Critical Concepts*, 474–492 London: Routledge.

Fotopoulos, Takis. 2000. Limitations of life-style strategies: The ecovillage movement is not the way toward a new democratic society. *Democracy and Nature* 6 (2): 287–308.

Global Ecovillage Network. 2004. http://www.gen.ecovillage.org.

Holmgren, David. 2002. *Permaculture: Principles and Pathways beyond Sustainability*. Victoria, Australia: Holmgren Design Services.

Holmgren, David, and Bill Mollison. 1978. *Permaculture One*. Australia: Transworld.

Jackson, Ross J. T. 2000. *We ARE Doing It: Building an Ecovillage Future*. San Francisco: Robert D. Reed Publishers.

Shellenberger, Michael and Nordhaus. Ted. 2004. *The Death of Environmentalism: Global Warming Politics in a Post-environmental World*. Oakland, CA: The Breakthrough Institute.

Speth, James Gustave. 2008. *The Bridge at the Edge of the World: Capitalism, the Environment, and Crossing from Crisis to Sustainability*. New Haven, CT: Yale University Press.

Snyder, Philip. 2006. Living and Learning Centers. http://gen.ecovillage.org/activities/living-learning/snydernew.php.

III

Philosophies of Localization

One trajectory of societal development has been toward the city-state and nation-state, toward empire and a balance of power. This is the one for which bold headlines and grand histories are written. Another trajectory, one that receives much less attention, has been toward villages and towns, ports and homesteads, farms and small shops. For simplicity, we might distinguish the two trajectories as *centralizing* and *decentralizing*.

Throughout history, it might seem that the decentralizing has always been overwhelmed by the centralizing. Future social chroniclers will assess whether this time, in the twenty-first century, it turns out different. We suspect it will. After all, never before has humanity faced global biophysical constraint. Never before have humans saturated the planet with human activity due to sheer size (seven billion and growing), extensive technologies (from earth movers to terminator genes), and high rates of consumption (for which another planet will be needed by 2030[1]).

What does seem clear, however, is that as empires rise and fall and powerful nations grow and then contract, the farmers, the yeomen, the small landholders, the shopkeepers, and the local manufacturers keep on going. These people provide the foundation—material, moral, and psychological—on which centralized power grows. But they provide more than just a stable base for this growth. They are not, as the following chapters will show, static forces maintaining the status quo. They are not about standing still or about regression. It is true that they are often portrayed, by those at society's center, as backward. But, often as not, they are sources of technological and cultural innovation and, from a sustainability perspective, they innovate largely in direct connection with the land and with each other.

So this section puts forward what we call *philosophies of localization*. The chapters offer historical, technological, ethical, and agrarian perspectives on, among other things, how decentralizing forces provide a stable societal base while, at the same time, providing a fount of fresh ideas and practices.

Notice that these contributions were written when the very idea of ecological overshoot, peak oil production, and climate disruption were, if anything, the wild imaginings of a disaffected few. Add to these authors' arguments the premise of this book and their arguments become all the more convincing.

Finally, note that none of the authors argue for complete decentralization; none say tear down the entire core and replace it with the periphery. Rather, they say that societies and their members will be better off—more democratic, community-oriented, self-regulating, and at home on this planet—if the *balance* between the trajectories shifts toward the decentralizing and away from the centralizing. Once again, with the premise of this book, we would only add that a biophysical imperative to quickly decentralize is now emerging and it will be better to guide it in positive directions.

Note

1. World Wide Fund for Nature (WWF International), Zoological Society of London, and Global Footprint Network, *Living Planet Report 2008* (Gland, Switzerland: WWF, 2008).

11

The Decentralist Tradition*

Kirkpatrick Sale

Kirkpatrick Sale, a writer and social critic, argues that a decentralist tradition is deeply rooted, ubiquitous, and continuous in human history. Showing humans' inclination toward local governance, this tradition of decentralization coexists with a centralizing tendency. However, neither is necessarily a response to the other nor depends on the other for its justification.

The implications for contemporary localization are several. One of the more important notions for the purposes of this book is that because the decentralist disposition is always present, it need not be created de novo. For instance, should central governments fail to respond in a timely fashion to the urgency of climate disruption and biophysical limits, local self-organizing will occur quite naturally, if not easily. At the same time, localizing need not be reactionary in that proponents need not wait for central failures to occur before they begin a shift toward the local. The worldwide local food movement is a case in point.

Notice that the premises of decentralization are consistent with the psychological premises of this book. Humans are problem-solving creatures. We have an inherited inclination to thrive within our given circumstances and we do so no matter how dire those circumstances may seem. Thus, applying Sale's work to localization, we suggest that the transition to a lower level of social complexity necessitated by emerging biophysical changes will be met by self-organization at a decentralized, local level. In contrast, centralized systems will only be able to respond by adding more social complexity, an effort that will indeed hinder more than help the transition (see Tainter, chapter 3, this volume).

*Sale, Kirkpatrick. 1980. "The Decentralist Tradition." In *Human Scale*, 443–454. New York: Perigee Books. Excerpted and reprinted with permission.

The impulse to local governance, to separatism and independence, to regional autonomy, seems an eternal one and well-nigh ineradicable. The long experience of nation-states—in Europe going back several centuries at least, in parts of Asia somewhat longer—has not destroyed that impulse, not in those countries, such as Britain, say, or the United States, where the state has grown to be most powerful and ubiquitous, not those places, such as Iran, where it has been most overreaching and oppressive. Indeed what is remarkable during these long years is how this decentralist tradition remains so resilient—so resilient that every time the power of the nation-state is broken, as during wars or rebellions, immediately there spring up a variety of decentralized organizations—in the neighborhoods, in the factories and offices, in the barracks and universities—that reinstitute government in local, popular, and anti-authoritarian forms.

The historical evidence is unmistakable on this point. In Paris in 1871 the collapse of the empire gave birth to the Paris Commune and its popular assemblies, while every neighborhood began its own committees of governance and defense and most of the business of the capital went on as usual, only with the workers themselves in charge: in Hannah Arendt's words, "a swift disintegration of the old power, the sudden loss of control over the means of violence, and, at the same time, the amazing formation of a new power structure which owed its existence to nothing but the organizational impulses of the people themselves."[1]

In Russia in 1905, and then again more sweepingly in 1917, industrial workers organized themselves into committees that took over factories and shops in practically every industry after the owners and bureaucrats had fled, and the real work of running most of the cities was done—until Bolshevik violence eventually put an end to them—by local *soviets*, popularly elected assemblies.

In Germany at the end of World War I, workers and soldiers in a number of cities organized themselves into local councils—*Räte*—in defiance of the Social Democratic regime in Berlin, demanding a new German constitution based on local autonomy through a nationwide *Rätesystem*, in Munich even establishing for a time a *Räterepublik* of Bavaria where, an *un*sympathetic observer noted, "every individual found his own sphere of action and could behold, as it were, with his own eyes his own contribution to the events of the day."

In Spain in 1936–37 the collapse of the national government was followed by the emergence of independent collective governments in hundreds of smaller towns, as we have seen, as well as in a number of the larger cities—Barcelona and Alcoy, particularly—where entire industries were run by self-management, municipal services such as the electric works and streetcar systems were operated by independent collectives, and political and economic affairs were in the hands of "technical administrative" committees elected within each industry.

In Hungary in 1956, the uprising that toppled the Communist regime led immediately, even in a country little used to popular government, to the formation of an extraordinary array of local councils, in neighborhoods and coffee houses, offices and factories, among writers and soldiers, students and—*mirabile dictu*—civil servants, indeed everywhere in the society, gradually coalescing within days into a rough network capable at least for a time of running the entire country.

In Iran in 1979, after the fall of the totalitarian government of the Shah, local institutions long suppressed suddenly appeared overnight, independent *ayatollahs* took control of their own provincial towns, neighborhood *komitehs* (even their name reminiscent of the Paris Commune) emerged to control their own territories in the cities, and at least four of the ancient minorities—the Baluchis in the southwest, the Kurds and Azerbaijanis in the northeast, and the Arabs of the Persian Gulf—asserted their independence with armed rebellions and demonstrations in the streets.

The same sort of thing, too, can be found in America after 1776, in France in 1789, in several European capitals in 1848, in many parts of Italy during the 1850s, in parts of occupied France and Italy in 1918–19, in Ireland throughout the civil war of the 1920s, in China in 1949, in Cuba in 1959, in Czechoslovakia in 1968, in Chile in 1970, in Portugal in 1974–75, in Angola and Mozambique at that same time, in fact as near as I can tell wherever and whenever a central government loses its hold (and before some new centralizing force, as often as not proclaiming itself revolutionary, takes over). Hannah Arendt has studied this phenomenon, noting with some wonderment how local councils and societies "make their appearance in every genuine revolution throughout the nineteenth and twentieth centuries. . . ."

Everywhere it seems to be the case that the absence of government does not lead to bewilderment and confusion and disorder, as might be imagined if all of government's claims for itself were true, but rather to a resurgence of locally based forms, most often democratically chosen and scrupulously responsive, that turn out to be quite capable of managing the complicated affairs of daily life for many months, occasionally years, until they are forcibly suppressed by some new centralist state less democratic and less responsive. They seem to be, as Arendt says, "spontaneous organs of the people," expressive of the natural human scale of politics and inheritors of the long tradition of decentralism.

It is striking to re-read history with eyes opened to the persistence of this tradition, because at once you begin to see the existence of the anti-authoritarian, independent, self-regulating, local community is every bit as basic to the human record as the existence of the centralized, imperial, hierarchical state, and far more ancient, more durable, and more widespread.

Obviously for the 3 million years that humanoids were becoming human they lived in small clans and groups, . . . and for the next 10,000 years that they were becoming "civilized" they lived in small communities and towns, needing none but the most limited kinds of governmental structures. Throughout the era of oriental empires—Persian, Sumerian, Egyptian, Babylonian—the greater part of the world's people still lived in independent hamlets, ever resistant to the imposition of outside authority, and even within the empires themselves local self-governing communes always persisted. Later, the Essenes, the people of the Dead Sea Scrolls who established an egalitarian community in opposition to Romanic Jerusalem in the second century before Christ, were only one of myriad tribes and sects that lived deliberately outside the Roman imperial influence. And still later the Christians themselves often lived in democratic and independent communities, sometimes in secret and sometimes openly but always apart from and hostile to whatever state might claim sovereignty.

The settlements of Greece were typical of such resistant localism: for many centuries they clung to a fierce independence, city upon city, valley after valley, no matter what putative conquerors might intrude, in time achieving that Hellenic civilization that is still a marvel of the world. . . .

Even to call it "Greece" is indeed to employ a modern fiction: the citizens of that ancient culture thought of themselves as Athenians and Spartans and Thebans, not Greeks, alike in language and civilization but not in political stamp or rule.

Traditional historians write of the European period from the fall of Rome to the Renaissance as if nothing much were going on outside of the consolidation of feudal families into the monarchies of the subsequent nation-states. But that is like talking of the night as the presence of stars or the ocean as if it were only waves. What was going on throughout the continent from the Atlantic to the Urals, what kept European civilization alive for better than ten centuries, was the maintenance and development of small, independent communities here in the form of Teutonic and Russian and Saxon villages with their popular councils and judicial elders, there as the medieval city-states with their guilds and brotherhoods and folkmotes, and over there as the chartered towns spread by the hundreds over France and Belgium with their special instruments of sovereignty and self-jurisdiction. Characteristic of the look of the continent were the divided cantons of what became Switzerland, beginning with the first democratic commune in Uri in the 1230s, a form that spread through dozens of villages in the fourteenth and fifteenth centuries and lasted until dominance by Napoleon at the end of the eighteenth century; at its height a typical canton, the independent Swiss Republic of the Three Leagues, covering about the area of Dallas, consisted of three loosely federated leagues, twenty-six sub-jurisdictions, forty-nine jurisdictional communes, and 227 autonomous neighborhoods—and, as an eighteenth-century traveler put it, "each village . . . each parish and each neighbourhood already constituted a tiny republic."

That Europe did eventually evolve some families designating themselves royal, and that some of those conquered vast areas of land they liked to call nations, and that the whole became a system of border-drawn nation-states such as we know today, does not mean that this was the tide and trend of that long era. Indeed, as between the statist tradition and the decentralist, these thousand years were clearly the period of the latter, into the fifteenth century in western European territories, in some places into the nineteenth century in eastern. No one has understood this period better than the Russian scientist and anarchist Peter Kropotkin, whose careful researches into its long-neglected intricacies, built upon the absolute explosion of interest in village government by scholars everywhere in the nineteenth century, have given us a telling picture of those centuries:

> Self-jurisdiction was the essential point [of the commune] and self-jurisdiction meant self-administration. . . . It had the right of war and peace, of federation and alliance with its neighbors. It was sovereign in its own affairs, and mixed with no others.
>
> In all its affairs the village commune was sovereign. Local custom was the law, and the plenary assembly of all the heads of family, men and women, was the judge, the only judge, in civil and criminal matters. . . .
>
> [In medieval towns] each street or parish had its popular assembly, its forum, its popular tribunal, its priest, its militia, its banner, and often its seal, the symbol of its sovereignty. Though federated with other streets it nevertheless maintained its independence.

This was the rule, mind, not the exception; it was exactly this self-governing community, through pestilence and war, the vicissitudes of nature and of kings, that sustained the many tens of millions of people of Europe for a millennium and more.

Nor did the tradition end with the rise of the nation-state. In many places it persisted quite a time—France did not outlaw local folkmotes until 1789, and Russian communes continued to exist in countless places until finally gutted by Stalin as late as the 1930s—and even in the age of nationalism it is not difficult to find, just below the surface, the roughly independent peasant village, the headstrong town, the self-minding neighborhood, in almost any country of Europe.

And in America. The decentralist tradition, manifested in a persistent anti-authoritarianism and a quite exuberant localism, is basic to the American character. (I am thinking of the European element here, but of course before that was the culture of the original Americans, almost everywhere communal, non-hierarchic, anti-institutional, and carefully localized.)

The Plymouth settlers, after all, were a proud and independent people who made the journey precisely to escape the press of the authoritarian state, and their original village was egalitarian enough, at least in its first two years, to have communal farming, the equal distribution of clothes and food, and cooking and laundry done in common. And when that first village tried to assert its control over such free spirits as Anne Hutchinson and Roger Williams, they simply moved on and started their own independent colonies—the beginning of a long, regular, native pattern of settlement that marked this land for at least three hundred years, until the Pacific stopped the march.

Others, too, who came here were anti-authoritarian by temperament, or tempering—the Quakers, and the Mennonites, escaping state persecution, the freed convicts and indentured workers, the entrepreneurs and political free-thinkers seeking fresh territory. Even such modest governments as the colonies represented seemed to chafe such people, and their resistance climaxed in the "insurrections" of 1675–90, in response to which William and Mary granted new and more lenient colonial charters. That did not halt the movement toward independence, however, and even the kinds of concessions later offered by George III and his ministers—and they were generous—did not succeed in abating the strong separatist spirit. Resistance to unwanted laws and the flouting of colonial authority were common well before the Revolution itself, and riots and rebellions—the Regulator movement against the governments of the Carolinas in the 1760s, for example, and the Green Mountain Boys against the government of New York in the 1770s—were recurrent. These fledgling Americans wanted to be left alone, to sink their roots how and where they pleased.

The Revolution was precisely in this tradition, and the document that began it is permeated with the principles of the sanctity of community borrowed from the philosophers of the Scottish Enlightenment, of the primacy of the people over the state plucked from Rousseau, and of the inviolability of local governance that was largely ingrained in the Americans themselves. No better confirmation of these principles was needed than the experience of the colonies in the early years after the Declaration, when most of the British institutions had collapsed and many of their leaders fled, and yet the citizens went right on administering their own affairs, and successfully too. Largely through town meetings, common from Massachusetts to Virginia and not alone in New England, the settlements of the new country raised money and volunteers for the new army (in which, incidentally, the soldiers elected their own officers), organized militias for self-defense, and took care of the plowing and planting, the road-repairing and bridge-building, the schooling and policing. As Tom Paine was later to write:

> For upwards of two years from the commencement of the American War, and for a longer period in several of the American states, there were no established forms of government. The old governments had been abolished and the country was too much occupied in defense to employ its attention in establishing new governments; yet during this interval order and harmony were preserved as inviolate as in any country of Europe.

The government that eventually did take shape over these lands, under the Articles of Confederation, was little more than an extension, a federation, of these existing forms. The Articles, much maligned by statists and regularly misconstrued by textbooks written from the viewpoint of a later age, were "weak" enough, as conventional opinion has it, if by "weak" is meant that the affairs of the country would continue to be the stuff of the daily chaffer-mugger of the village square and the town meeting and not matters exclusively for professionals in some inaccessible capital; "weak" if by weak is meant that, in the words of the Articles' first and most basal provision, "each state retains sovereignty, freedom, and independence"; "weak" if popular government be weak, if local control be weak, if direct democracy be weak. Such matters will always be murky, but there is excellent evidence that the greatest part of the free population supported the Articles wholeheartedly and was little interested in the drive for a stronger government that such people as Hamilton and Madison began pushing after the war. And even when the centralists and commercial interests pushed it through, the Constitution was approved only after the state legislators, "in order to prevent misconstruction or abuse of its powers," demanded that a Bill of Rights be added to it: the danger of the central government . . . was uppermost even in the minds of those who were constructing it. . . .

Not that the youthful United States government was a particularly autocratic one, not that at all. In fact it was run by men who saw themselves as truly libertarian, in service to those principles of federalism and republicanism that did much to spread power out from the center to the state capitals and the counties and towns. But it did not take more than a few dozen years before the [astute] Thomas Jefferson, who had done so much to assure that the new government would be restrained, began to fear that even this much centralism was beginning to rot the republic and remove the essential affairs of state from those who should of right be tangling with them. Around 1816, after having served his stint to the presidency perhaps not wisely nor too well, he began to revive an idea that had long been part of his political creed: ward government. A system of small "elementary republics," he began to feel—units of perhaps a hundred men or two, populations of 500 to 1,000 in all—was essential to the salvation of the American state, and a better alternative than his earlier notion of recurring revolutions ("a little rebellion, now and then"). What he urged on all who would hear him was "small republics" by which "every man in the State" could become "an acting member of the Common government, transacting in person a great portion of its rights and duties, subordinate indeed, yet important, and entirely within his competence." "Divide [government] among the many," he declared, so that each citizen may feel "that he is a participator in the government of affairs, not merely at an election one day in the year, but every day; when there shall not be a man in the State who will not be a member of some one of its councils, great or small, he will let the heart be torn out of his body sooner than his power wrested from him by a Caesar or a Bonaparte." Thus a nation of face-to-face democracy, of town meetings, of neighborhood government, where "the voice of the whole people would be fairly, fully, and peaceably expressed, discussed, and decided by the common reason" of every citizen, every day, everywhere.

The Jeffersonian formula was never tried, not even seriously debated in the land—Jefferson himself had effectively retired from public life and chose not to enter new lists at this late date, his passions not quite up to his convictions. And even as it was being voiced, the large shadow of the Federal and state government was moving slowly out to dim and extinguish the small lights of self-government that existed: one after another the towns and cities of the mid-Atlantic region abandoned town meetings or made them ritualistic annual affairs; the corporate form of city government, with mayors and councils, was pushed by conservatives as a way to keep the peace and put decisions into "responsible" hands; and the township system was downgraded in favor of greater power to the state and, to a somewhat lesser degree, Federal governments. It was, as historian Merrill Jensen has put it in his authoritative *The New Nation*, "the counter-revolution."

But the Jeffersonian ideal remained, in one form or another, the philosophic pole at one end of American politics throughout at least the first half of the nineteenth century. It was the guide for Thoreau ("That government is best which governs not at all") and for Emerson ("The less government we have, the better—the fewer laws, and the less confided power"), for Calhoun and the Carolina Nullificationists, and for both white abolitionists and black insurrectionaries who repeatedly defied state and Federal laws. At the time of the Civil War there were many who would say with Thoreau:

> This government never of itself furthered any enterprise, but by the alacrity with which it got out of the way. *It* does not keep the country free. *It* does not settle the West. *It* does not educate. The character inherent in the American people has done all that has been accomplished; and it would have done somewhat more, if the government had not sometimes got in its way.

The Civil War and its centralizing aftermath—wars are always centralizing; that's why governments have them—brought a temporary halt to this tradition, but it broke out again with real ferocity once more in the latter part of the century. Against the increasing monolithicity of government, industry, and political party, there sprang up a variety of movements diverse in cause but similar in resistance to the centralists: the Greenbacks, the Grangers, Oklahoma Socialists, Knights of Labor, Georgists, feminists, anarchists, communists, utopians, and above all the groups that fused to become, around 1880, the Populists. The Populists seemed almost to be that party Emerson had dreamed of —"fanatics in freedom: they hated tolls, taxes, turnpikes, banks, hierarchies, governors, yea, almost laws"—except that they added to this anti-authoritarianism a profound regard for communalism, cooperation, federation, networking, and localism, and actually developed an extraordinary variety of ventures to foster those. From Texas to the Carolinas, the Populists represented a major part of American politics, winning over whole towns in the South and West, achieving electoral victories in several states—in fact, gaining control of the North Carolina legislature in 1890 and passing laws for local self-government through county autonomy. It may in the end have proved a failure, but Populism was American to the grain, built upon the small farmers and artisans of the still-frontier settlements, rooted in the values of the local community and those enterprises—grange, church, school, newspaper, local shops—that gave them expression, set against all those Eastern institutions—industrial trusts, railroads, big-city machines, national banks—that in fact in time were to suffocate those enterprises.

With the first two decades of the twentieth century the triumph of Federal power was made manifest. The central government was acknowledged as supreme, its authority over its population's pockets (the Income Tax Amendment of 1913) and habits (the Prohibition Amendment of 1919) and even lives (the Selective Service Act of 1917) fully established. Those who resisted, on whatever grounds, were given a show of raw Federal power: the Espionage Act of 1917, the Immigration Acts of 1917–18, the Sedition Act of 1918, the Red Raids of 1919, the Palmer Raids of 1920, and countless little sins of commission in between. What happened then in the 1930s and 1940s, with the familiar events of New Deal consolidation, seemed only a natural extension of the past.

Even then, the decentralist spirit did not disappear. It found expression in the Agrarian movement of the 1920s and 1930s against corporate giantism and the growing industrialization and urbanization of the South; in the cooperative movement, both agricultural and consumer, that set roots then that are still in place today; and in a variety of homegrown radicalisms of Ralph Borsodi's homesteaders, Arthur Morgan's communitarians, the Catholic distributists, the Technocrats, the folkschoolers, the Black Mountain anarchists, and so on. Many of these eventually joined to found the journal *Free America*, which for ten difficult years from 1937 to 1946 represented, as nothing in this century has, the voice of decentralism in America, summed up in its founding creed:

> *Free America* stands for individual independence and believes that freedom can exist only in societies in which the great majority are the effective owners of

tangible and productive property and in which group action is democratic. In order to achieve such a society, ownership, production, population, and government must be decentralized. *Free America* is therefore opposed to finance-capitalism, fascism, and communism.

. . . The decentralist tradition, no matter what, will not die, for it is as wide in the American soul as the country is wide, as deep in the American psyche as the riches are deep. One may well wonder, with historian C. Vann Woodward, why, given its steady opposition to centralization and authority, the American environment nonetheless "should have proved so hospitable to those same tendencies in government, military, and business. A huge federal bureaucracy," he taunts, "a great military establishment, and multinational corporations, not to mention big labor, seem to have successfully surmounted all the handicaps to centralization." But the answer is as easy as it is revealing. The centralizing tendency has always existed in this country alongside the decentralizing—for every Anne Hutchinson a Governor Winthrop, for every Jefferson a Hamilton, for every Calhoun a Webster, for every Thoreau a Longfellow, for every Debs a Wilson, for every Borsodi a Tugwell, for every Brandeis a Frankfurter, for every Mumford a Schlesinger, for every Schumacher a Galbraith. And obviously this century, not only in this country but around the world, has belonged to the centralists and all their totalitarian machinery. But the decentralist movement, if it has not triumphed of late, has not disappeared either, and it seems to have survived Woodward's decades of bureaucracies and multinationals quite intact. Indeed, one gets the sense that these next few decades may provide its chance again; and more: that these decades offer the opportunity for it to establish its patterns—of localism, self-sufficiency, ecological harmony, and participatory democracy—for a long time to come. . . .

Note

1. Karl Marx himself saw in this neighborhood government system "the political form of even the smallest village," which looked as if it might be "the political form, finally discovered, for the economic liberation of labor." He was right, of course, but he forgot it.

12

Technology at a Human Scale*

Ernst F. Schumacher

Economist Ernst F. Schumacher wrote what can be considered an outline for durable living where an acute sensitivity to scale, both organizational and technological, is central. In Small Is Beautiful *he sought to integrate meaningful, moral, and conscious living into a model of economics more noble than the dominant neoclassical model.*

In this chapter, Schumacher suggests that technology, which often promises a liberation from work and a leisurely existence, as practiced actually saps real living from people. The question then becomes, is there a technology that focuses not just on increasing efficiency, but on promoting proficiency, the ability to self-provision and self-govern? Notice that Schumacher's answer is to propose a different kind of technology, one with a human face. Particularly striking is how he debunks the notion that anyone who critiques the dominant forms of technology is anti-progress. In this book, we take such critiques as essential to human adaptation in the face of unavoidable decline in available material and energy, and the concomitant increase in available time and quality of life.

The modern world has been shaped by its metaphysics, which has shaped its education, which in turn has brought forth its science and technology. So, without going back to metaphysics and education, we can say that the modern world has been shaped by technology. It tumbles from crisis to crisis; on all sides there are prophecies of disaster and, indeed, visible signs of breakdown.

*Schumacher, Ernst F. 1995. "Technology with a human face." Excerpted from pp. 146–159 of *Small Is Beautiful: Economics as if People Mattered*. Published by Harper & Row and Blond & Briggs and Hutchinson.

If that which has been shaped by technology, and continues to be so shaped, looks sick, it might be wise to have a look at technology itself. If technology is felt to be becoming more and more inhuman, we might do well to consider whether it is possible to have something better—a technology with a human face.

Strange to say, technology, although of course the product of man, tends to develop by its own laws and principles, and these are very different from those of human nature or of living nature in general. Nature always, so to speak, knows where and when to stop. Greater even than the mystery of natural growth is the mystery of the natural cessation of growth. There is measure in all natural things—in their size, speed, or violence. As a result, the system of nature, of which man is a part, tends to be self-balancing, self-adjusting, self-cleansing. Not so with technology, or perhaps I should say: not so with man dominated by technology and specialisation. Technology recognises no self-limiting principle—in terms, for instance, of size, speed, or violence. It therefore does not possess the virtues of being self-balancing, self-adjusting, and self-cleansing. In the subtle system of nature, technology, and in particular the super-technology of the modern world, acts like a foreign body, and there are now numerous signs of rejection.

Suddenly, if not altogether surprisingly, the modern world, shaped by modern technology, finds itself involved in three crises simultaneously. First, human nature revolts against inhuman technological, organisational, and political patterns, which it experiences as suffocating and debilitating; second, the living environment which supports human life aches and groans and gives signs of partial breakdown; and, third, it is clear to anyone fully knowledgeable in the subject matter that the inroads being made into the world's non-renewable resources, particularly those of fossil fuels, are such that serious bottlenecks and virtual exhaustion loom ahead in the quite foreseeable future.

Any one of these three crises or illnesses can turn out to be deadly. I do not know which of the three is the most likely to be the direct cause of collapse. What is quite clear is that a way of life that bases itself on materialism, i.e., on permanent, limitless expansionism in a finite environment, cannot last long, and that its life expectation is the shorter the more successfully it pursues its expansionist objectives. . . .

The primary task of technology, it would seem, is to lighten the burden of work man has to carry in order to stay alive and develop his potential. It is easy enough to see that technology fulfills this purpose when we watch any particular piece of machinery at work—a computer, for instance, can do in seconds what it would take clerks or even mathematicians a very long time, if they can do it at all. It is more difficult to convince oneself of the truth of this simple proposition when one looks at whole societies. When I first began to travel the world, visiting rich and poor countries alike, I was tempted to formulate the first law of economics as follows: "The amount of real leisure a society enjoys tends to be in inverse proportion to the amount of labour-saving machinery it employs." It might be a good idea for the professors of economics to put this proposition into their examination papers and ask their pupils to discuss it. However that may be, the evidence is very strong indeed. If you go from easy-going England to, say, Germany or the United States, you find that people there live under much more strain than here. And if you move to a country like Burma, which is very near to the bottom of the league table of industrial progress, you find that people have an enormous amount of leisure really to enjoy themselves. Of course, as there is so much less labour-saving machinery to help them, they "accomplish" much less than we do; but that is a different point. The fact remains that the burden of living rests much more lightly on their shoulders than on ours.

The question of what technology actually does for us is therefore worthy of investigation. It obviously greatly reduces some kinds of work while it increases other kinds. The type of work which modern technology is most successful in reducing or even eliminating is skilful, productive work of human hands, in touch with real materials of one kind or another. In an advanced industrial society, such work has become exceedingly rare, and to make a decent living by doing such work has become virtually impossible. A great part of the modern neurosis may be due to this very fact; for the human being, defined by Thomas Aquinas as a being with brains and hands, enjoys nothing more than to be creatively, usefully, productively engaged with both his hands and his brains. Today, a person has to be wealthy to be able to enjoy this simple thing, this very great luxury: he has to be able to afford space and good tools; he has to be lucky enough to find a good teacher and plenty of free time to learn and practise. He really has to be rich enough not to need a job; for the number of jobs that would be satisfactory in these respects is very small indeed.

The extent to which modern technology has taken over the work of human hands may be illustrated as follows. We may ask how much of "total social time"—that is to say, the time all of us have together, twenty-four hours a day each—is actually engaged in real production.

Rather less than one-half of the total population of this country is, as they say, gainfully occupied, and about one-third of these are actual producers in agriculture, mining, construction, and industry. I do mean *actual producers*, not people who tell other people what to do, or account for the past, or plan for the future, or distribute what other people have produced. In other words, rather less than one-sixth of the total population is engaged in actual production; on average, each of them supports five others beside himself; of which two are gainfully employed on things other than real production and three are not gainfully employed. Now, a fully employed person, allowing for holidays, sickness, and other absence, spends about one-fifth of his total time on his job. It follows that the proportion of "total social time" spent on actual production—in the narrow sense in which I am using the term—is, roughly, one-fifth of one-third of one-half; i.e., 3½ per cent. The other 96½ per cent of "total social time" is spent in other ways, including sleeping, eating, watching television, doing jobs that are not *directly* productive, or just killing time more or less humanely.

Although this bit of figuring work need not be taken too literally, it quite adequately serves to show what technology has enabled us to do: namely, to reduce the amount of time actually spent on production in its most elementary sense to such a tiny percentage of total social time that it pales into insignificance, that it carries no real weight, let alone prestige. When you look at industrial society in this way, you cannot be surprised to find that prestige is carried by those who help fill the other 96½ per cent of total social time, primarily the entertainers but also the executors of Parkinson's Law. In fact, one might put the following proposition to students of sociology: "The prestige carried by people in modern industrial society varies in inverse proportion to their closeness to actual production."

There is a further reason for this. The process of confining productive time to 3½ per cent of total social time has had the inevitable effect of taking all normal human pleasure and satisfaction out of the time spent on this work. Virtually all real production has been turned into an inhuman chore which does not enrich a man but empties him. "From the factory," it has been said, "dead matter goes out improved, whereas men there are corrupted and degraded."

We may say, therefore, that modern technology has deprived man of the kind of work that he enjoys most, creative, useful work with hands and brains, and given him plenty of work of a fragmented kind, most of which he does not enjoy at all. It has multiplied the number of people who are exceedingly busy doing kinds of work which, if it is productive at all, is so only in an indirect or "roundabout" way, and much of which would not be necessary at all if technology were rather less modern. . . .

Taking stock, we can say that we possess a vast accumulation of new knowledge, splendid scientific techniques to increase it further, and immense experience in its application. All this is truth of a kind. This truthful knowledge, as such, does *not* commit us to a technology of giantism, supersonic speed, violence, and the destruction of human work-enjoyment. The use we have made of our knowledge is only one of its possible uses and, as is now becoming ever more apparent, often an unwise and destructive use.

As I have shown, directly productive time in our society has already been reduced to about 3½ per cent of total social time, and the whole drift of modern technological development is to reduce it further, asymptotically to zero. Imagine we set ourselves a goal in the opposite direction—to increase it sixfold, to about twenty per cent, so that twenty per cent of total social time would be used for actually producing things, employing hands and brains and, naturally, excellent tools. An incredible thought! Even children would be allowed to make themselves useful, even old people. At one-sixth of present-day productivity, we should be producing as much as at present. There would be six times as much time for any piece of work we chose to undertake—enough to make a really good job of it, to enjoy oneself, to produce real quality, even to make things beautiful. Think of the therapeutic value of real work; think of its educational value. No one would then want to raise the school-leaving age or to lower the retirement age, so as to keep people off the labour market. Everybody would be welcome to lend a hand. Everybody would be admitted to what is now the rarest privilege, the opportunity of working usefully, creatively, with his own hands and brains, in his own time, at his own pace—and with excellent tools. Would this mean an enormous extension of working hours? No, people who work in this way do not know the difference between work and leisure. Unless they sleep or eat or occasionally choose to do nothing at all, they are always agreeably, productively engaged. Many of the "on-cost jobs" would simply disappear; I leave it to the reader's imagination to identify them. There would be little need for mindless entertainment or other drugs, and unquestionably much less illness.

Now, it might be said that this is a romantic, a utopian, vision. True enough. What we have today, in modern industrial society, is not romantic and certainly not utopian, as we have it right here. But it is in very deep trouble and holds no promise of survival. We jolly well have to have the courage to dream if we want to survive and give our children a chance of survival. The threefold crisis of which I have spoken will not go away if we simply carry on as before. It will become worse and end in disaster, until or unless we develop a new life-style which is compatible with the real needs of human nature, with the health of living nature around us, and with the resource endowment of the world.

Now, this is indeed a tall order, not because a new life-style to meet these critical requirements and facts is impossible to conceive, but because the present consumer society is like a drug addict who, no matter how miserable he may feel, finds it extremely difficult to get off the hook. The problem children of the world—from this point of view and in spite of many other considerations that could be adduced—are the rich societies and not the poor. . . .

As Gandhi said, the poor of the world cannot be helped by mass production, only by production by the masses. The system of *mass production*, based on sophisticated, highly capital-intensive, high energy-input dependent, and human labour-saving technology, presupposes that you are already rich, for a great deal of capital investment is needed to establish one single workplace. The system of *production by the masses* mobilises the priceless resources which are possessed by all human beings, their clever brains and skilful hands, *and supports them with first-class tools*. The technology of mass production is inherently violent, ecologically damaging, self-defeating in terms of non-renewable resources, and stultifying for the human person. The technology of *production by the masses*, making use of the best of modern knowledge and experience, is conducive to decentralisation, compatible with the laws of ecology, gentle in its use of scarce resources, and designed to serve the human person instead of making him the servant of machines. I have named it *intermediate technology* to signify that it is vastly superior to the primitive technology of bygone ages but at the same time much simpler, cheaper, and freer than the supertechnology of the rich. One can also call it self-help technology, or democratic or people's technology—a technology to which everybody can gain admittance and which is not reserved to those already rich and powerful. . . .

Although we are in possession of all requisite knowledge, it still requires a systematic, creative effort to bring this technology into active existence and make it generally visible and available. It is my experience that it is rather more difficult to recapture directness and simplicity than to advance in the direction of ever more sophistication and complexity. Any third-rate engineer or researcher can increase complexity; but it takes a certain flair of real insight to make things simple again. And this insight does not come easily to people who have allowed themselves to become alienated from real, productive work and from the self-balancing system of nature, which never fails to recognise measure and limitation. Any activity which fails to recognise a self-limiting principle is of the devil. In our work with the developing countries we are at least forced to recognise the limitations of poverty, and this work can therefore be a wholesome school for all of us in which, while genuinely trying to help others, we may also gain knowledge and experience of how to help ourselves.

I think we can already see the conflict of attitudes which will decide our future. On the one side, I see the people who think they can cope with our threefold crisis by the methods current, only more so; I call them the people of the forward stampede. On the other side, there are people in search of a new life-style, who seek to return to certain basic truths about man and his world; I call them home-comers. Let us admit that the people of the forward stampede, like the devil, have all the best tunes or at least the most popular and familiar tunes. You cannot stand still, they say; standing still means going down; you must go forward; there is nothing wrong with modern technology except that it is as yet incomplete; let us complete it. Dr. Sicco Mansholt, one of the most prominent chiefs of the European Economic Community, may be quoted as a typical representative of this group. "More, further, quicker, richer," he says, "are the watchwords of present-day society." And he thinks we must help people to adapt, "for there is no alternative." This is the authentic voice of the forward stampede, which talks in much the same tone as Dostoyevsky's Grand Inquisitor: "Why have you come to hinder us?" They point to the population explosion and to the possibilities of world hunger. Surely, we must take our flight forward and not be faint-hearted. If people start protesting and revolting, we shall have to have more police and have them better equipped. If there is trouble with the environment, we shall need more stringent laws against pollution, and faster economic growth to pay for antipollution measures. If there are problems about natural resources, we shall turn to synthetics; if there are problems about fossil fuels, we shall move from slow reactors to fast breeders and from fission to fusion. There *are* no insoluble problems.

The slogans of the people of the forward stampede burst into the newspaper headlines every day with the message, "a breakthrough a day keeps the crisis at bay."

And what about the other side? This is made up of people who are deeply convinced that technological development has taken a wrong turn and needs to be redirected. The term "home-comer" has, of course, a religious connotation. For it takes a good deal of courage to say "no" to the fashions and fascinations of the age and to question the presuppositions of a civilisation which appears destined to conquer the whole world; the requisite strength can be derived only from deep convictions. If it were derived from nothing more than fear of the future, it would be likely to disappear at the decisive moment. . . .

The home-comers base themselves upon a different picture of man from that which motivates the people of the forward stampede. It would be very superficial to say that the latter believe in "growth" while the former do not. In a sense, everybody believes in growth, and rightly so, because growth is an essential feature of life. The whole point, however, is to give to the idea of growth a qualitative determination; for there are always many things that ought to be growing and many things that ought to be diminishing.

Equally, it would be very superficial to say that the home-comers do not believe in progress, which also can be said to be an essential feature of all life. The whole point is to determine what constitutes progress. And the home-comers believe that the direction which modern technology has taken and is continuing to pursue—towards ever-greater size, ever-higher speeds, and ever-increased violence, in defiance of all laws of natural harmony—is the opposite of progress. Hence the call for taking stock and finding a new orientation. The stocktaking indicates that we are destroying our very basis of existence, and the reorientation is based on remembering what human life is really about.

In one way or another everybody will have to take sides in this great conflict. To "leave it to the experts" means to side with the people of the forward stampede. It is widely accepted that politics is too important a matter to be left to experts. Today, the main content of politics is economics, and the main content of economics is technology. If politics cannot be left to the experts, neither can economics and technology.

The case for hope rests on the fact that ordinary people are often able to take a wider view, and a more "humanistic" view, than is normally being taken by experts. The power of ordinary people, who today tend to feel utterly powerless, does not lie in starting new lines of action, but in placing their sympathy and support with minority groups which have already started. I shall give two examples relevant to the subject here under discussion. One relates to agriculture, still the greatest single activity of man on earth, and the other relates to industrial technology.

Modern agriculture relies on applying to soil, plants, and animals ever-increasing quantities of chemical products, the long-term effect of which on soil fertility and health is subject to very grave doubts. People who raise such doubts are generally confronted with the assertion that the choice lies between "poison or hunger." There are highly successful farmers in many countries who obtain excellent yields without resort to such chemicals and without raising any doubts about long-term soil fertility and health. For the last twenty-five years, a private, voluntary organisation, the Soil Association, has been engaged in exploring the vital relationships between soil, plant, animal, and man; has undertaken and assisted relevant research; and has attempted to keep the public informed about developments in these fields. Neither the successful farmers nor the Soil Association have been able to attract official support or recognition. They have generally been dismissed as "the muck and mystery people," because they are obviously outside the mainstream of modern technological progress. Their methods bear the mark of non-violence and humility towards the infinitely subtle system of natural harmony, and this stands in opposition to the life-style of the modern world. But if we now realise that the modern life-style is putting us into mortal danger, we may find it in our hearts to support and even join these pioneers rather than to ignore or ridicule them.

On the industrial side, there is the Intermediate Technology Development Group. It is engaged in the systematic study on how to help people to help themselves. While its work is primarily concerned with giving technical assistance to the Third World, the results of its research are attracting increasing attention also from those who are concerned about the future of the rich societies. For they show that an intermediate technology, a technology with a human face, is in fact possible; that it is viable; and that it reintegrates the human being, with his skilful hands and creative brain, into the productive process. It serves *production by the masses* instead of *mass production*. Like the Soil Association, it is a private, voluntary organisation depending on public support.

I have no doubt that it is possible to give a new direction to technological development, a direction that shall lead it back to the real needs of man, and that also means: *to the actual size of man*. Man is small, and, therefore, small is beautiful. To go for giantism is to go for self-destruction. And what is the cost of a reorientation? We might remind ourselves that to calculate the cost of survival is perverse. No doubt, a price has to be paid for anything worthwhile: to redirect technology so that it serves man instead of destroying him requires primarily an effort of the imagination and an abandonment of fear.

13

Provincialism*

Josiah Royce

Writing more than a century ago, philosopher Josiah Royce attempts to rehabilitate a maligned term: provincialism. *That he did not succeed is less important than the fact that, in his time, he saw the need for a social process that counters and complements nationalization. His provincialism is roughly equivalent to what Hess in chapter 21 of this book calls* localism *and what we put under the larger notion of localization.*

Significantly, Royce does not see provincialism as retreat, as a means of insular living, or even as staying small. Rather, provincialism gives identity and meaning to individuals, and innovation and social solidarity to the nation. Notice how for Royce, the local and the national go hand in hand. Also notice how Royce insists that provincialism has no specified geographic or political scale, only that it is less than the nation and has its own identity.

Finally, his purpose is to show that a "Higher Provincialism" can act as a "saving power to which the world in the near future will need more and more to appeal." Unable to forecast the rise of our present globalized commercial order, its consumerist culture, the dominance of oil and the automobile, let alone two world wars, the Great Depression, global environmental threats and biophysical limits, he may have thought that that future was only years away, not decades, let alone an entire century. Under the premises of this book, Royce's future may be now.

I propose, in this address, to define certain issues which, as I think, the present state of the world's civilization, and of our own national life, make both prominent and critical.

*Royce, Josiah. 1908. *Race Questions: Provincialism and Other American Problems.* New York: The Macmillan Company.

I

The word "provincialism," which I have used as my title, has been chosen because it is the best single word that I have been able to find to suggest the group of social tendencies to which I want to call your especial attention. I intend to use this word in a somewhat elastic sense, which I may at once indicate. When we employ the word "provincialism" as a concrete term, speaking of "a provincialism," we mean, I suppose, any social disposition, or custom, or form of speech or of civilization, which is especially characteristic of a province. In this sense one speaks of the provincialisms of the local dialect of any English shire, or of any German country district. This use of the term in relation to the dialects of any language is very common. But one may also apply the term to name, not only the peculiarities of a local dialect, but the fashions, the manners, and customs of a given restricted region of any country. One also often employs the word "provincialism" as an abstract term, to name not only the customs or social tendencies themselves, but that fondness for them, that pride in them, which may make the inhabitants of a province indisposed to conform to the ways of those who come from without, and anxious to follow persistently their own local traditions. Thus the word "provincialism" applies both to the social habits of a given region, and to the mental interest which inspires and maintains these habits. But both uses of the term imply, of course, that one first knows what is to be meant by the word "province." This word, however, is one of an especially elastic usage. Sometimes, by a province, we mean a region as restricted as a single English county, or as the smallest of the old German principalities. Sometimes, however, one speaks of the whole of New England, or even of the Southern states of our Union, as constituting one province; and I know of no easy way of defining how large a province may be. For the term, in this looser sense, stands for no determinate political or legal division of a country. Meanwhile we all, in our minds, oppose the term "province" to the term "nation," as the part is opposed to the whole. Yet we also often oppose the terms "provincial" and "metropolitan," conceiving that the country districts and the smaller towns and cities belong even to the province, while the very great cities belong rather to the whole country, or even to the world in general. Yet here the distinction that we make is not the same as the former distinction between the part of a country and the whole country. Nevertheless, the ground for such an identification of the provincial with that which pertains to country districts and to smaller cities can only lie in the supposed tendency of the great city to represent better the interests of the larger whole than do the lesser communities. This supposition, however, is certainly not altogether well founded. In the sense of possessing local interests and customs, and of being limited to ideas of their own, many great cities are almost as distinctly provincial as are certain less populous regions. The plain people of London or of Berlin have their local dialect; and it seems fair to speak of the peculiarities of such dialects as provincialisms. And almost the same holds true of the other social traditions peculiar to individual great cities. It is possible to find, even amongst the highly cultivated classes of ancient cities, ideas and fashions of behavior as characteristically local, as exclusive in their indifference to the ways of outsiders, as are the similarly characteristic ways and opinions of the country districts of the same nationality. And so the opposition of the provincial to the metropolitan, in manners and in beliefs, seems to me much less important than the other opposition of the province, as the more or less restricted part, to the nation as the whole. It is this latter opposition that I shall therefore emphasize in the present discussion. But I shall not attempt to define how large or how well organized, politically, a province must be. For my present purpose a county, a state, or even a large section of the country, such as New England, might constitute a province. For me, then, a province shall mean any one part of a national domain, which is geographically and socially sufficiently unified to have a true consciousness of its own unity, to feel a pride in its own ideals and customs, and to possess a sense of its distinction from other parts of the country. And by the term "provincialism" I shall mean, first, the tendency of such a province to possess its own customs and ideals; secondly, the totality of these customs and ideals themselves; and thirdly, the love and pride which lead the inhabitants of a province to cherish as their own these traditions, beliefs, and aspirations.

II

I have defined the term used as my title. But now, in what sense do I propose to make provincialism our topic? You will foresee that I intend to discuss the worth of provincialism, i.e., to consider, to some extent, whether it constitutes a good or an evil element in civilization. You will properly expect me, therefore, to compare provincialism with other social tendencies; such tendencies as patriotism, the larger love of humanity, and the ideals of higher cultivation. Precisely these will constitute, in fact, the special topics of my address. But all that I have to say will group itself about a single thesis, which I shall forthwith announce. My thesis is that, in the present state of the world's civilization, and of the life of our own country, the time has come to emphasize, with a new meaning and intensity, the positive value, the absolute necessity for our welfare, of a wholesome provincialism, as a saving power to which the world in the near future will need more and more to appeal.

The time was (and not very long since), when, in our own country, we had to contend against very grave evils due to false forms of provincialism. What has been called sectionalism long threatened our national unity. Our Civil War was fought to overcome the ills due to such influences. There was, therefore, a time when the virtue of true patriotism had to be founded upon a vigorous condemnation of certain powerful forms of provincialism. And our national education at that time depended both upon our learning common federal ideals, and upon our looking to foreign lands for the spiritual guidance of older civilizations. Furthermore, not only have these things been so in the past, but similar needs will, of course, be felt in the future. We shall always be required to take counsel of the other nations in company with whom we are at work upon the tasks of civilization. Nor have we outgrown our spiritual dependence upon older forms of civilization. In fact we shall never outgrow a certain inevitable degree of such dependence. Our national unity, moreover, will always require of us a devotion that will transcend in some directions the limits of all our provincial ideas. A common sympathy between the different sections of our country will, in future, need a constantly fresh cultivation. Against the evil forms of sectionalism we shall always have to contend. All this I well know, and these things I need not in your presence emphasize. But what I am to emphasize is this: The present state of civilization, both in the world at large, and with us, in America, is such as to define a new social mission which the province alone, but not the nation, is able to fulfill. False sectionalism, which disunites, will indeed always remain as great an evil as ever it was. But the modern world has reached a point where it needs, more than ever before, the vigorous development of a highly organized provincial life. Such a life, if wisely guided, will not mean disloyalty to the nation; and it need not mean narrowness of spirit, nor yet the further development of jealousies between various communities. What it will mean, or at least may mean,—this, so far as I have time, I wish to set forth in the following discussion. My main intention is to define the right form and the true office of provincialism,—to portray what, if you please, we may well call the Higher Provincialism,—to portray it, and then to defend it, to extol it, and to counsel you to further just such provincialism.

Since this is my purpose, let me at once say that I address myself, in the most explicit terms, to men and women who, as I hope and presuppose, are and wish to be, in the wholesome sense, provincial. Every one, as I maintain, ought, ideally speaking, to be provincial,—and that no matter how cultivated, or humanitarian, or universal in purpose or in experience he may be or may become. If in our own country, where often so many people are still comparative strangers to the communities in which they have come to live, there are some of us who, like myself, have changed our provinces during our adult years, and who have so been unable to become and to remain in the sense of European countries provincial; and if, moreover, the life of our American provinces everywhere has still too brief a tradition,—all that is our misfortune, and not our advantage. As our country grows in social organization, there will be, in absolute measure, more and not less provincialism amongst our people. To be sure, as I hope, there will also be, in absolute measure, more and not less patriotism, closer and not looser national ties, less and not more mutual sectional misunderstanding. But the two tendencies, the tendency toward national unity and that toward local independence of spirit, must henceforth grow together. They cannot prosper apart. The national unity must not kill out, nor yet hinder, the provincial self-consciousness. The loyalty to the Republic must not lessen the love and the local pride of the individual community. The man of the future must love his province more than he does to-day. His provincial customs and ideals must be more and not less highly developed, more and not less self-conscious, well-established, and earnest. And therefore, I say, I appeal to you as to a company of people who are, and who mean to be, provincial as well as patriotic,—servants and lovers of your own community and of its ways, as well as citizens of the world. I hope and believe that you all intend to have your community live its own life, and not the life of any other community, nor yet the life of a mere abstraction called humanity in general. I hope that you are fully aware how provincialism, like monogamy, is an essential basis of true civilization. And it is with this presupposition that I undertake to suggest something toward a definition and defense of the higher provincialism and of its office in civilization.

III

With this programme in mind, let me first tell you what seem to me to be in our modern world, and, in particular, in our American world, the principal evils which are to be corrected by a further development of a true provincial spirit, and which cannot be corrected without such a development. The first of these evils I have already mentioned. It is a defect incidental, partly to the newness of our own country, but partly also to those world-wide conditions of modern life which make travel, and even a change of home, both attractive and easy to dwellers in the most various parts of the globe. In nearly every one of our American communities, at least in the northern and in the western regions of our country, there is a rather large proportion of people who either have not grown up where they were born, or who have changed their dwelling-place in adult years. I can speak all the more freely regarding this class of our communities, because, in my own community, I myself, as a native of California, now resident in New England, belong to such a class. Such classes, even in modern New England, are too large. The stranger, the sojourner, the newcomer, is an inevitable factor in the life of most American communities. To make him welcome is one of the most gracious of the tasks in which our people have become expert. To give him his fair chance is the rule of our national life. But it is not on the whole well when the affairs of a community remain too largely under the influence of those who mainly feel either the wanderer's or the new resident's interest in the region where they are now dwelling. To offset the social tendencies due to such frequent changes of dwelling-place we need the further development and the intensification of the community spirit. The sooner the new resident learns to share this spirit, the better for him and for his community. A sound instinct, therefore, guides even our newer communities, in the more fortunate cases, to a rapid development of such a local sentiment as makes the stranger feel that he must in due measure conform if he would be permanently welcome, and must accept the local spirit if he is to enjoy the advantages of his community. As a Californian I have been interested to see both the evidences and the nature of this rapid evolution of the genuine provincial spirit in my own state. How swiftly, in that country, the Californians of the early days seized upon every suggestion that could give a sense of the unique importance of their new provincial life. The associations that soon clustered about the tales of the life of Spanish missionaries and Mexican colonists in the years before 1846,—these our American Californians cherished from the outset. This, to us often half-legendary past, gave us a history of our own. The wondrous events of the early mining life,—how earnestly the pioneers later loved to rehearse that story; and how proud every young Californian soon became of the fact that his father had had his part therein. Even the Californian's well-known and largely justified glorification of his climate was, in his own mind, part of the same expression of his tendency to idealize whatever tended to make his community, and all its affairs, seem unique, beloved, and deeply founded upon some significant natural basis. Such a foundation was, indeed, actually there; nature had, indeed, richly blessed his land; but the real interest that made one emphasize and idealize all these things, often so boastfully, was the interest of the loyal citizen in finding his community an object of pride. . . .

IV

This second modern evil arises from, and constitutes, one aspect of the leveling tendency of recent civilization. That such a leveling tendency exists, most of us recognize. That it is the office of the province to contend against some of the attendant evils of this tendency, we less often observe. By the leveling tendency in question I mean that aspect of modern civilization which is most obviously suggested by the fact that, because of the ease of communication amongst distant places, because of the spread of popular education, and because of the consolidation and of the centralization of industries and of social authorities, we tend all over the nation, and, in some degree, even throughout the civilized world, to read the same daily news, to share the same general ideas, to submit to the same overmastering social forces, to live in the same external fashions, to discourage individuality, and to approach a dead level of harassed mediocrity. One of the most marked of all social tendencies is in any age that toward the mutual assimilation of men in so far as they are in social relations with one another. One of the strongest human predispositions is that toward imitation. But our modern conditions have greatly favored the increase of the numbers of people who read the same books and newspapers, who repeat the same phrases, who follow the same social fashions, and who thus, in general, imitate one another in constantly more and more ways. The result is a tendency to crush the individual. Furthermore there are modern economic and industrial developments, too well known to all of you to need any detailed mention here, which lead toward similar results. The independence of the small trader or manufacturer becomes lost in the great commercial or industrial combination. The vast corporation succeeds and displaces the individual. Ingenuity and initiative become subordinated to the discipline of an impersonal social order. And each man, becoming, like his fellow, the servant of masters too powerful for him to resist, and too complex in their undertakings for him to understand, is, in so far, disposed unobtrusively to conform to the ways of his innumerable fellow-servants, and to lose all sense of his unique moral destiny as an individual. . . .

And in similar fashion provincial pride helps the individual man to keep his self-respect even when the vast forces that work toward industrial consolidation, and toward the effacement of individual initiative, are besetting his life at every turn. For a man is in large measure what his social consciousness makes him. Give him the local community that he loves and cherishes, that he is proud to honor and to serve, make his ideal of that community lofty,—give him faith in the dignity of his province,—and you have given him a power to counteract the leveling tendencies of modern civilization.

V

The third of the evils with which a wise provincialism must contend is closely connected with the second. I have spoken of the constant tendency of modern life to the mutual assimilation of various parts of the social order. Now this assimilation may occur slowly and steadily, as in great measure it normally does; or, on the other hand, it may take more sudden and striking forms, at moments when the popular mind is excited, when great emotions affect the social order. At such times of emotional disturbance, society is subject to tendencies which have recently received a good deal of psychological study. They are the tendencies to constitute what has often been called the spirit of the crowd or of the mob. Modern readers of the well-known book of Le Bon's on "The Crowd" well know what the tendencies to which I refer may accomplish. It is true that the results of Le Bon are by no means wholly acceptable. It is true that the psychology of large social masses is still insufficiently understood, and that a great many hasty statements have been made about the fatal tendency of great companies of people to go wrong. Yet in the complex world of social processes there can be no doubt that there exist such processes as the ones which Le Bon characterizes. The mob-spirit is a genuine psychological fact which occasionally becomes important in the life of all numerous communities. Moreover, the mob-spirit is no new thing. It has existed in some measure from the very beginning of social life. But there are certain modem conditions which tend to give the mob-spirit new form and power, and to lead to new social dangers that are consequent upon the presence of this spirit. . . .

Now these principles are responsible for the explanation of the well-known contrast between those social phenomena which illustrate the wisdom of the enlightened social order, and the phenomena which, on the contrary, often seem such as to make us despair for the moment of the permanent success of popular government. In the rightly constituted social group where every member feels his own responsibility for his part of the social enterprise which is in hand, the result of the interaction of individuals is that the social group may show itself wiser than any of its individuals. In the mere crowd, on the other hand, the social group may be, and generally is, more stupid than any of its individual members. Compare a really successful town meeting in a comparatively small community with the accidental and sometimes dangerous social phenomena of a street mob or of a great political convention. In the one case every individual may gain wisdom from his contact with the social group. In the other case every man concerned, if ever he comes again to himself, may feel ashamed of the absurdity of which the whole company was guilty. Social phenomena of the type that may result from the higher social group, the group in which individuality is respected, even while social loyalty is demanded,—these phenomena may lead to permanent social results which as tradition gives them a fixed character may gradually lead to the formation of permanent institutions, in which a wisdom much higher than that of any individual man may get embodied. . . .

There are social groups that are not subject to the mob-spirit. And now if you ask how such social groups are nowadays to be fostered, to be trained, to be kept alive for the service of the nation, I answer that the place for fostering such groups is the province, for such groups flourish under conditions that arouse local pride, the loyalty to one's own community, the willingness to remember one's own ways and ideals, even at the moment when the nation is carried away by some leveling emotion. The lesson would then be: Keep the province awake, that the nation may be saved from the disastrous hypnotic slumber so characteristic of excited masses of mankind.

VI

I have now reviewed three types of evils against which I think it is the office of provincialism to contend. As I review these evils, I am reminded somewhat of the famous words of Schiller in his "Greeting to the New Century," which he composed at the outset of the nineteenth century. In his age, which in some respects was so analogous to our own, despite certain vast differences, Schiller found himself overwhelmed as he contemplated the social problem of the moment by the vast national conflict, and the overwhelming forces which seemed to him to be crushing the more ideal life of his nation, and of humanity. With a poetic despair that we need indeed no longer share, Schiller counsels his reader, in certain famous lines, to flee from the stress of life into the still recesses of the heart, for, as he says, beauty lives only in song, and freedom has departed into the realm of dreams. Now Schiller spoke in the romantic period. We no longer intend to flee from our social ills to any realm of dreams. And as to the recesses of the heart, we now remember that out of the heart, are the issues of life. But so much my own thesis and my own counsel would share in common with Schiller's words. I should say to-day that our national unities have grown so vast, our forces of social consolidation have become so paramount, the resulting problems, conflicts, evils, have been so intensified, that we, too, must flee in the pursuit of the ideal to a new realm. Only this realm is, to my mind, so long as we are speaking of social problems, a realm of real life. It is the realm of the province. There must we flee from the stress of the now too vast and problematic life of the nation as a whole. There we must flee. I mean, not in the sense of a cowardly and permanent retirement, but in the sense of a search for renewed strength, for a social inspiration, for the salvation of the individual from the overwhelming forces of consolidation. Freedom, I should say, dwells now in the small social group, and has its securest home in the provincial life. The nation by itself, apart from the influence of the province, is in danger of becoming an incomprehensible monster, in whose presence the individual loses his right, his self-consciousness, and his dignity. The province must save the individual. But, you may ask, in what way do I conceive that the wise provincialism of which I speak ought to undertake and carry on its task? How is it to meet the evils of which I have been speaking? In what way is its influence to be exerted against them? And how can the province cultivate its self-consciousness without tending to fall back again into the ancient narrowness from which small communities were so long struggling to escape? How can we keep broad humanity and yet cultivate provincialism? How can we be loyally patriotic, and yet preserve our consciousness of the peculiar and unique dignity of our own community? In what form are our wholesome provincial activities to be carried on?

I answer, of course, in general terms, that the problem of the wholesome provincial consciousness is closely allied to the problem of any individual form of activity. An individual tends to become narrow when he is what we call self-centered. But, on the other hand, philanthropy that is not founded upon a personal loyalty of the individual to his own family and to his own personal duties is notoriously a worthless abstraction. We love the world better when we cherish our own friends the more faithfully. We do not grow in grace by forgetting individual duties in behalf of remote social enterprises. Precisely so, the province will not serve the nation best by forgetting itself, but by loyally emphasizing its own duty to the nation and therefore its right to attain and to cultivate its own unique wisdom. Now all this is indeed obvious enough, but this is precisely what in our days of vast social consolidation we are some of us tending to forget.

Now as to the more concrete means whereby the wholesome provincialism is to be cultivated and encouraged, let me appeal directly to the loyal member of any provincial community, be it the community of a small town, or of a great city, or of a country district. Let me point out what kind of work is needed in order to cultivate that wise provincialism which, as you see, I wish to have grow not in opposition to the interests of the nation, but for the very sake of saving the nation from the modern evil tendencies of which I have spoken.

First, then, I should say a wholesome provincialism is founded upon the thought that while local pride is indeed a praiseworthy accompaniment of every form of social activity, our province, like our own individuality, ought to be to all of us rather an ideal than a mere boast. And here, as I think, is a matter which is too often forgotten. Everything valuable is, in our present human life, known to us as an ideal before it becomes an attainment, and in view of our human imperfections, remains to the end of our short lives much more a hope and an inspiration than it becomes a present achievement. Just because the true issues of human life are brought to a finish not in time but in eternity, it is necessary that in our temporal existence what is most worthy should appear to us as an ideal, as an Ought, rather than as something that is already in our hands. The old saying about the bird in the hand being worth two in the bush does not rightly apply to the ideal goods of a moral agent working under human limitations. For him the very value of life includes the fact that its goal as something infinite can never at any one instant be attained. In this fact the moral agent glories, for it means that he has something to do. Hence the ideal in the bush, so to speak, is always worth infinitely more to him than the food or the plaything of time that happens to be just now in his hands. The difference between vanity and self-respect depends largely upon this emphasizing of ideals in the case of the higher forms of self-consciousness, as opposed to the emphasis upon transient temporal attainments in the case of the lower forms. Now what holds true of individual self-consciousness, ought to hold true of the self-consciousness of the community. Boasting is often indeed harmless and may prove a stimulus to good work. It is therefore to be indulged as a tribute to our human weakness. But the better aspect of our provincial consciousness is always its longing for the improvement of the community.

And now, in the second place, a wise provincialism remembers that it is one thing to seek to make ideal values in some unique sense our own, and it is quite another thing to believe that if they are our own, other people cannot possess such ideal values in their own equally unique fashion. A realm of genuinely spiritual individuality is one where each individual has his own unique significance, so that none could take another's place. But for just that very reason all the unique individuals of the truly spiritual order stand in relation to the same universal light, to the same divine whole in relation to which they win their individuality. Hence all the individuals of the true spiritual order have ideal goods in common, as the very means whereby they can win each his individual place with reference to the possession and the employment of these common goods. Well, it is with provinces as with individuals. The way to win independence is by learning freely from abroad, but by then insisting upon our own interpretation of the common good. A generation ago the Japanese seemed to most European observers to be entering upon a career of total self-surrender. They seemed to be adopting without stint European customs and ideals. They seemed to be abandoning their own national independence of spirit. They appeared to be purely imitative in their main purposes. They asked other nations where the skill of modern sciences lay, and how the new powers were to be gained by them. They seemed to accept with the utmost docility every lesson, and to abandon with unexampled submissiveness, their purpose to remain themselves. Yet those of us who have watched them since, or who have become acquainted with representative Japanese students, know how utterly superficial and illusory that old impression of ours was regarding the dependence, or the extreme imitativeness, or the helpless docility, of the modern Japanese. He has now taught us quite another lesson. With a curious and on the whole not unjust spiritual wiliness, he has learned indeed our lesson, but he has given it his own interpretation. You always feel in intercourse with a Japanese how unconquerable the spirit of his nation is, how inaccessible the recesses of his spirit have remained after all these years of free intercourse with Europeans. In your presence the Japanese always remains the courteous and respectful learner so long as he has reason to think that you have anything to teach him. But he remains as absolutely his own master with regard to the interpretation, the use, the possession of all spiritual gifts, as if he were the master and you the learner. He accepts the gifts, but their place in his national and individual life is his own. And we now begin to see that the feature of the Japanese nationality as a member of the civilized company of nations is to be something quite unique and independent. Well, let the Japanese give us a lesson in the spirit of true provincialism. Provincialism does not mean a lack of plasticity, an unteachable spirit; it means a determination to use the spiritual gifts that come to us from abroad in our own way and with reference to the ideals of our own social order.

And therefore, thirdly, I say in developing your provincial spirit, be quite willing to encourage your young men to have relations with other communities. But on the other hand, encourage them also to make use of what they thus acquire for the furtherance of the life of their own community. Let them win aid from abroad, but let them also have, so far as possible, an opportunity to use this which they acquire in the service of their home. Of course economic conditions rather than deliberate choice commonly determine how far the youth of a province are able to remain for their lifetime in a place where they grow up. But so far as a provincial spirit is concerned, it is well to avoid each of two extremes in the treatment of the young men of the community,—extremes that I have too often seen exemplified. The one extreme consists in maintaining that if young men mean to be loyal to their own province, to their own state, to their own home, they ought to show their loyalty by an unwillingness to seek guidance from foreign literature, from foreign lands, in the patronizing of foreign or distant institutions, or in the acceptance of the customs and ideas of other communities than their own. Against this extreme let the Japanese be our typical instance. They have wandered far. They have studied abroad. They have assimilated the lore of other communities. And they have only gained in local consciousness, in independence of spirit, by the ordeal. The other extreme is the one expressed in that tendency to wander and to encourage wandering, which has led so many of our communities to drive away the best and most active of their young men. We want more of the determination to find, if possible, a place for our youth in their own communities. Finally, let the province more and more seek its own adornment. Here I speak of a matter that in all our American communities has been until recently far too much neglected. Local pride ought above all to centre, so far as its material objects are concerned, about the determination to give the surroundings of the community nobility, dignity, beauty. We Americans spend far too much of our early strength and time in our newer communities upon injuring our landscapes, and far too little upon endeavoring to beautify our towns and cities. We have begun to change all that, and while I have no right to speak as an aesthetic judge concerning the growth of the love of the beautiful in our country, I can strongly insist that no community can think any creation of genuine beauty and dignity in its public buildings or in the surroundings of its towns and cities too good a thing for its own deserts. For we deserve what in such realms we can learn how to create or to enjoy, or to make sacrifices for. And no provincialism will become dangerously narrow so long as it is constantly accompanied by a willingness to sacrifice much in order to put in the form of great institutions, of noble architecture, and of beautiful surroundings an expression of the worth that the community attaches to its own ideals.

14

Local Enterprise*

Wendell Berry

In this chapter, writer and farmer Wendell Berry characterizes the productive units of a localized society. Among the defining features are independent ownership grounded in a physical place, long-term maintenance of that place, local knowledge, and good work. Notice that while Berry focuses on the family farm, he is really talking about all productive enterprises that are both "the home and the workplace" of those who own the enterprise. The true value of such enterprises, he argues, accrues not only to the owners but to the broader community and society. His, in short, is an ethic of practice in place, *not of economic growth, consumption, or industrial values.*

Defending the family farm is like defending the Bill of Rights or the Sermon on the Mount or Shakespeare's plays. One is amazed at the necessity for defense, and yet one agrees gladly, knowing that the family farm is both eminently defensible and a part of the definition of one's own humanity. But having agreed to this defense, one remembers uneasily that there has been a public clamor in defense of the family farm throughout all the years of its decline—that, in fact, "the family farm" has become a political catchword, like democracy and Christianity, and much evil has been done in its name.

Several careful distinctions are therefore necessary. What I shall mean by the term "family farm" is a farm small enough to be farmed by a family and one that *is* farmed by a family, perhaps with a small amount of hired help. I shall *not* mean a farm that is owned by a family and worked by other people. The family farm is both the home and the workplace of the family that owns it.

By the verb "farm," I do not mean just the production of marketable crops but also the responsible maintenance of the health and usability of the place while it is in production. A family farm is one that is properly cared for by its family.

Furthermore, the term "family farm" implies longevity in the connection between family and farm. A family farm is not a farm that a family has bought on speculation and is only occupying and using until it can be profitably sold. Neither, strictly speaking, is it a farm that a family has newly bought, though, depending on the intentions of the family, we may be able to say that such a farm is *potentially* a family farm. This suggests that we may have to think in terms of ranks or degrees of family farms. A farm that has been in the same family for three generations may rank higher as a family farm than a farm that has been in a family only one generation; it may have a higher degree of familyness or familiarity than the one-generation farm. Such distinctions have a practical usefulness to the understanding of agriculture, and, as I hope to show, there are rewards of longevity that do not accrue only to the farm family.

I mentioned the possibility that a family farm might use a small amount of hired help. This greatly complicates matters, and I wish it were possible to say, simply, that a family farm is farmed with family labor. But it seems important to allow for the possibility of supplementing family labor with wagework or some form of sharecropping. Not only may family labor become insufficient as a result, say, of age or debility but also an equitable system of wage earning or sharecropping would permit unpropertied families to earn their way to farm ownership. The critical points, in defining "family farm," are that the amount of nonfamily labor should be small and that it should supplement, not replace, family labor. On a family farm, the family members are workers, not overseers. If a family on a family farm does require supplementary labor, it seems desirable that the hired help should live on the place and work year-round; the idea of a family farm is jeopardized by supposing that the farm family might be simply the guardians or maintainers of crops planted and harvested by seasonal workers. These requirements, of course, imply both small scale and diversity.

Finally, I think we must allow for the possibility that a family farm might be very small or marginal and that it might not entirely support its family. In such cases, though the economic return might be reduced, the *values* of the family-owned and family-worked small farm are still available both to the family and to the nation.

The idea of the family farm, as I have just defined it, is conformable in every way to the idea of good farming—that is, farming that does not destroy either farmland or farm people. The two ideas may, in fact, be inseparable. If family farming and good farming are as nearly synonymous as I suspect they are, that is because of a law that is well understood, still, by most farmers but that has been ignored in the colleges, offices, and corporations of agriculture for thirty-five or forty years. The law reads something like this: Land that is in human use must be lovingly used; it requires intimate knowledge, attention, and care.

The practical meaning of this law (to borrow an insight from Wes Jackson[1]) is that there is a ratio between eyes and acres, between farm size and farm hands, that is correct. We know that this law is unrelenting—that, for example, one of the meanings of our current high rates of soil erosion is that we do not have enough farmers; we have enough farmers to use the land but not enough to use it and protect it at the same time.

In this law, which is not subject to human repeal, is the justification of the small, family-owned, family-worked farm, for this law gives a preeminent and irrevocable value to familiarity, the family life that alone can properly connect a people to a land. This connection, admittedly, is easy to sentimentalize, and we must be careful not to do so. We all know that small family farms can be abused because we know that sometimes they have been; nevertheless, it is true that familiarity tends to mitigate and to correct abuse. A family that has farmed land through two or three generations will possess not just the land but a remembered history of its own mistakes and of the remedies of those mistakes. It will know, not just what it *can* do, what is technologically possible, but also what it *must* do and what it must *not* do; the family will have understood the ways in which it and the farm empower and limit one another. This is the value of longevity in landholding: In the long term, knowledge and affection accumulate, and, in the long term, knowledge and affection pay. They do not just pay the family in goods and money; they also pay the family and the whole country in health and satisfaction.

But the justifications of the family farm are not merely agricultural; they are political and cultural as well. The question of the survival of the family farm and the farm family is one version of the question of who will own the country, which is, ultimately, the question of who will own the people. Shall the usable property of our country be democratically divided, or not? Shall the power of property be a democratic power, or not? If many people do not own the usable property, then they must submit to the few who do own it. They cannot eat or be sheltered or clothed except in submission. They will find themselves entirely dependent on money; they will find costs always higher, and money always harder to get. To renounce the principle of democratic property, which is the only basis of democratic liberty, in exchange for specious notions of efficiency or the economics of the so-called free market is a tragic folly.

There is one more justification, among many, that I want to talk about—namely, that the small farm of a good farmer, like the small shop of a good craftsman or craftswoman, gives work a quality and a dignity that it is dangerous, both to the worker and the nation, for human work to go without. If using ten workers to make one pin results in the production of many more pins than the ten workers could produce individually, that is undeniably an improvement in production, and perhaps uniformity is a virtue in pins. But, in the process, ten workers have been demeaned; they have been denied the economic use of their minds; their work has become thoughtless and skill-less. Robert Heilbroner says that such "division of labor reduces the activity of labor to dismembered gestures."[2]

. . . Eric Gill sees in this industrial dismemberment of labor a crucial distinction between *making* and *doing*, and he describes "the degradation of the mind" that is the result of the shift from making to doing.[3] This degradation of the mind cannot, of course, be without consequences. One obvious consequence is the degradation of products. When workers' minds are degraded by loss of responsibility for what is being made, they cannot use judgment; they have no use for their critical faculties; they have no occasions for the exercise of workmanship, of workmanly pride. And the consumer is degraded by loss of the opportunity for qualitative choice. This is why we must now buy our clothes and immediately resew the buttons; it is why our expensive purchases quickly become junk.

With industrialization has come a general depreciation of work. As the price of work has gone up, the value of it has gone down, until it is now so depressed that people simply do not want to do it anymore. We can say without exaggeration that the present national ambition of the United States is unemployment. People live for quitting time, for weekends, for vacations, and for retirement; moreover, this ambition seems to be classless, as true in the executive suites as on the assembly lines. One works, not because the work is necessary, valuable, useful to a desirable end, or because one loves to do it, but only to be able to quit—a condition that a saner time would regard as infernal, a condemnation. This is explained, of course, by the dullness of the work, by the loss of responsibility for, or credit for, or knowledge of the thing made. What can be the status of the working small farmer in a nation whose motto is a sigh of relief: "Thank God it's Friday"?

But there is an even more important consequence: By the dismemberment of work, by the degradation of our minds as workers, we are denied our highest calling, for, as Gill says, "every man is called to give love to the work of his hands. Every man is called to be an artist."[4] The small family farm is one of the last places—they are getting rarer every day—where men and women (and girls and boys, too) can answer that call to be an artist, to learn to give love to the work of their hands. It is one of the last places where the maker—and some farmers still do talk about "making the crops"—is responsible, from start to finish, for the thing made. This certainly is a spiritual value, but it is not for that reason an impractical or uneconomic one. In fact, from the exercise of this responsibility, this giving of love to the work of the hands, the farmer, the farm, the consumer, and the nation all stand to gain in the most practical ways: They gain the means of life, the goodness of food, and the longevity and dependability of the sources of food, both natural and cultural. The proper answer to the spiritual calling becomes, in turn, the proper fulfillment of physical need.

The family farm, then, is good, and to show that it is good is easy. Those who have done most to destroy it have, I think, found no evil in it. But, if a good thing is failing among us, pretty much without being argued against and pretty much without professed enemies, then we must ask why it should fail. I have spent years trying to answer this question, and, while I am sure of some answers, I am also sure that the complete answer will be hard to come by because the complete answer has to do with who and what we are as a people; the fault lies in our identity and therefore will be hard for us to see.

However, we must *try* to see, and the best place to begin may be with the fact that the family farm is not the only good thing that is failing among us. The family farm is failing because it belongs to an order of values and a kind of life that are failing. We can only find it wonderful, when we put our minds to it, that many people now seem willing to mount an emergency effort to "save the family farm" who have not yet thought to save the family or the community, the neighborhood schools or the small local businesses, the domestic arts of household and homestead, or cultural and moral tradition—all of which are also failing, and on all of which the survival of the family farm depends.

The family farm is failing because the pattern it belongs to is failing, and the principal reason for this failure is the universal adoption, by our people and our leaders alike, of industrial values, which are based on three assumptions:

1. That value equals price—that the value of a farm, for example, is whatever it would bring on sale, because both a place and its price are "assets." There is no essential difference between farming and selling a farm.
2. That all relations are mechanical. That a farm, for example, can be used like a factory, because there is no essential difference between a farm and a factory.
3. That the sufficient and definitive human motive is competitiveness—that a community, for example, can be treated like a resource or a market, because there is no difference between a community and a resource or a market.

The industrial mind is a mind without compunction; it simply accepts that people, ultimately, will be treated as things and that things, ultimately, will be treated as garbage.

Such a mind is indifferent to the connections, which are necessarily both practical and cultural, between people and land; which is to say that it is indifferent to the fundamental economy and economics of human life. Our economy is increasingly abstract, increasingly a thing of paper, unable either to describe or to serve the real economy that determines whether or not people will eat and be clothed and sheltered. And it is this increasingly false or fantastical economy that is invoked as a standard of national health and happiness by our political leaders.

That this so-called economy can be used as a universal standard can only mean that it is itself without standards. Industrial economists cannot measure the economy by the health of nature, for they regard nature as simply a source of "raw materials." They cannot measure it by the health of people, for they regard people as "labor" (that is, as tools or machine parts) or as "consumers." They can measure the health of the economy only in sums of money.

Here we come to the heart of the matter—the absolute divorce that the industrial economy has achieved between itself and all ideals and standards outside itself. It does this, of course, by arrogating to itself the status of primary reality. Once that is established, all its ties to principles of morality, religion, or government necessarily fall slack.

But a culture disintegrates when its economy disconnects from its government, morality, and religion. If we are dismembered in our economic life, how can we be members in our communal and spiritual life? We assume that we can have an exploitive, ruthlessly competitive, profit-for-profit's-sake economy, and yet remain a decent and a democratic nation, as we still apparently wish to think ourselves. This simply means that our highest principles and standards have no practical force or influence and are reduced merely to talk.

That this is true was acknowledged by William Safire in a recent column, in which he declared that our economy is driven by greed and that greed, therefore, should no longer count as one of the seven deadly sins. "Greed," he said, "is finally being recognized as a virtue . . . the best engine of betterment known to man." It is, moreover, an agricultural virtue: "The cure for world hunger is the driving force of Greed." Such statements would be possible only to someone who sees the industrial economy as the ultimate reality. Mr. Safire attempts a disclaimer, perhaps to maintain his status as a conservative: "I hold no brief for Anger, Envy, Lust, Gluttony, Pride, Envy or Sloth."[5] But this is not a cat that can be let only partly out of the bag. In fact, all seven of the deadly sins are "driving forces" of this economy, as its advertisements and commercials plainly show.

As a nation, then, we are not very religious and not very democratic, and *that* is why we have been destroying the family farm for the last forty years—along with other small local economic enterprises of all kinds. We have been willing for millions of people to be condemned to failure and dispossession by the workings of an economy utterly indifferent to any claims they may have had either as children of God or as citizens of a democracy. "That's the way a dynamic economy works," we have said. We have said, "Get big or get out." We have said, "Adapt or die." And we have washed our hands of them.

Throughout this period of drastic attrition on the farm, we supposedly have been "subsidizing agriculture," but, as Wes Jackson has pointed out,[6] this is a misstatement. What we have actually been doing is using the farmers to launder money for the agribusiness corporations, which have controlled both their supplies and their markets, while the farmers have overproduced and been at the mercy of the markets. The result has been that the farmers have failed by the millions, and the agribusiness corporations have prospered—or they prospered until the present farm depression, when some of them have finally realized that, after all, they are dependent on their customers, the farmers.

Throughout this same desperate time, the colleges of agriculture, the experiment stations, and the extension services have been working under their old mandate to promote "a sound and prosperous agriculture and rural life," to "aid in maintaining an equitable balance between agriculture and other segments of the economy," to contribute "to the establishment and maintenance of a permanent and effective agricultural industry," and to help "the development and improvement of the rural home and rural life."[7]

That the land-grant system has failed this commission is, by now, obvious. I am aware that there are many individual professors, scientists, and extension workers whose lives have been dedicated to the fulfillment of this commission and whose work has genuinely served the rural home and rural life. But, in general, it can no longer be denied that the system as a whole has failed. One hundred and twenty-four years after the Morrill Act, ninety-nine years after the Hatch Act, seventy-two years after the Smith-Lever Act, the "industrial classes" are not liberally educated, agriculture and rural life are not sound or prosperous or permanent, and there is no equitable balance between agriculture and other segments of the economy. Anybody's statistics on the reduction of the farm population, on the decay of rural communities, on soil erosion, soil and water pollution, water shortages, and farm bankruptcies tell indisputably a story of failure.

This failure cannot be understood apart from the complex allegiances between the land-grant system and the aims, ambitions, and values of the agribusiness corporations. The willingness of land-grant professors, scientists, and extension experts to serve as state-paid researchers and traveling salesmen for those corporations has been well documented and is widely known.

The reasons for this state of affairs, again, are complex. I have already given some of them; I don't pretend to know them all. But I would like to mention one that I think is probably the most telling: that the offices of the land-grant complex, like the offices of the agricultural bureaucracy, have been looked upon by their aspirants and their occupants as a means, not to serve farmers, but to escape farming. Over and over again, one hears the specialists and experts of agriculture introduced as "old farm boys" who have gone on (as is invariably implied) to better things. The reason for this is plain enough: The life of a farmer has characteristically been a fairly hard one, and the life of a college professor or professional expert has characteristically been fairly easy. Farmers—working family farmers—do not have tenure, business hours, free weekends, paid vacations, sabbaticals, and retirement funds; they do not have professional status.

The direction of the career of agricultural professionals is, typically, not toward farming or toward association with farmers. It is "upward" through the hierarchy of a university, a bureau, or an agribusiness corporation. They do not, like Cincinnatus, leave the plow to serve their people and return to the plow. They leave the plow, simply, for the sake of leaving the plow.

This means that there has been for several decades a radical disconnection between the land-grant institutions and the farms, and this disconnection has left the land-grant professionals free to give bad advice; indeed, if they can get this advice published in the right place, from the standpoint of their careers it does not matter whether their advice is good or not.

For example, after years of milk glut, when dairy farmers are everywhere threatened by their surplus production, university experts are still working to increase milk production and still advising farmers to cull their least-productive cows—apparently oblivious both of the possible existence of other standards of judgment and of the fact that this culling of the least-productive cows is, ultimately, the culling of the smaller farmers.

Perhaps this could be dismissed as human frailty or inevitable bureaucratic blundering—except that the result is damage, caused by people who probably would not have given such advice if they were themselves in a position to suffer from it. Serious responsibilities are undertaken by public givers of advice, and serious wrong is done when the advice is bad. Surely a kind of monstrosity is involved when tenured professors with protected incomes recommend or even tolerate Darwinian economic policies for farmers, or announce (as one university economist after another has done) that the failure of so-called inefficient farmers is good for agriculture and good for the country. They see no inconsistency, apparently, between their own protectionist economy and the "free market" economy that they recommend to their supposed constituents, to whom the "free market" has proved, time and again, to be fatal. Nor do they see any inconsistency, apparently, between the economy of a university, whose sources, like those of any tax-supported institution, are highly diversified, and the extremely specialized economies that they have recommended to their farmer-constituents. These inconsistencies nevertheless exist, and they explain why, so far, there has been no epidemic of bankruptcies among professors of agricultural economics.

These, of course, are simply instances of the notorious discrepancy between theory and practice. But this discrepancy need not exist, or it need not be so extreme, in the colleges of agriculture. The answer to the problem is simply that those who profess should practice. Or at least a significant percentage of them should. This is, in fact, the rule in other colleges and departments of the university. A professor of medicine who was no doctor would readily be seen as an oddity; so would a law professor who could not try a case; so would a professor of architecture who could not design a building. What, then, would be so strange about an agriculture professor who would be, and who would be expected to be, a proven farmer?

But it would be wrong, I think, to imply that the farmers are merely the victims of their predicament and share none of the blame. In fact, they, along with all the rest of us, do share the blame, and their first hope of survival is in understanding that they do.

Farmers, as much as any other group, have subscribed to the industrial fantasies that I listed earlier: that value equals price, that all relations are mechanical, and that competitiveness is a proper and sufficient motive. Farmers, like the rest of us, have assumed, under the tutelage of people with things to sell, that selfishness and extravagance are merely normal. Like the rest of us farmers have believed that they might safely live a life prescribed by the advertisers of products, rather than the life required by fundamental human necessities and responsibilities.

One could argue that the great breakthrough of industrial agriculture occurred when most farmers became convinced that it would be better to own a neighbor's farm than to have a neighbor and when they became willing, necessarily at the same time, to borrow extravagant amounts of money. They thus violated the two fundamental laws of domestic or community economy: You must be thrifty and you must be generous; or, to put it in a more practical way, you must be (within reason) independent, and you must be neighborly. With that violation, farmers became vulnerable to everything that has intended their ruin.

An economic program that encourages the unlimited growth of individual holdings not only anticipates but actively proposes the failure of many people. Indeed, as our antimonopoly laws testify, it proposes the failure, ultimately, of all but one. It is a fact, I believe, that many people have now lost their farms and are out of farming who would still be in place had they been willing for their neighbors to survive along with themselves. In light of this, we see that the machines, chemicals, and credit that farmers have been persuaded to use as "labor savers" have, in fact, performed as neighbor replacers. And whereas neighborhood tends to work as a service free to its members, the machines, chemicals, and credit have come at a cost set by people who were *not* neighbors.

That is a description of the problem of the family farm, as I see it. It is a dangerous problem, but I do not think it is hopeless. On the contrary, a number of solutions to the problem are implied in my description of it.

What, then, can be done?

The most obvious, the most desirable, solution would be to secure that "equitable balance between agriculture and other segments of the economy" that is one of the stated goals of the Hatch Act. To avoid the intricacies of the idea of "parity," which we inevitably think of here, I will just say that the price of farm products, as they leave the farm, should be on a par with the price of those products that the farmer must buy.

In order to achieve this with minimal public expense, we must control agricultural production; supply must be adjusted to demand. Obviously this is something that individual farmers, or individual states, cannot do for themselves; it is a job that belongs appropriately to the federal government. As a governmental function, it is perfectly in keeping with the ideal, everywhere implicit in the originating documents of our government, that the small have a right to certain protections from the great. We have, within limits that are obvious and reasonable, the *right* to be small farmers or small businessmen or -women, just as, or perhaps insofar as, we have a right to life, liberty, and property. The individual citizen is not to be victimized by the rich any more than by the powerful. When Marty Strange writes, "To the extent that only the exceptional succeed, the system fails,"[8] he is economically and agriculturally sound, but he is also speaking directly from American political tradition.

The plight of the family farm would be improved also by other governmental changes—for example, in policies having to do with taxation and credit.

Our political problem, of course, is that farmers are neither numerous enough nor rich enough to be optimistic about government help. The government tends, rather, to find their surplus production useful and their economic failure ideologically desirable. Thus, it seems to me that we must concentrate on those things that farmers and farming communities can do for themselves—striving in the meantime for policies that would be desirable.

It may be that the gravest danger to farmers is their inclination to look to the government for help, after the agribusiness corporations and the universities (to which they have already looked) have failed them. In the process, they have forgotten how to look to themselves, to their farms, to their families, to their neighbors, and to their tradition.

Marty Strange has written also of his belief "that commercial agriculture can survive within pluralistic American society, as we know it—*if* [my emphasis] the farm is rebuilt on some of the values with which it is popularly associated: conservation, independence, self-reliance, family, and community. To sustain itself, commercial agriculture will have to reorganize its social and economic structure as well as its technological base and production methods in a way that reinforces these values."[9] I agree. Those are the values that offer us survival, not just as farmers, but as human beings. And I would point out that the transformation that Marty is proposing cannot be accomplished by the governments, the corporations, or the universities; if it is to be done, the farmers themselves, their families, and their neighbors will have to do it.

What I am proposing, in short, is that farmers find their way out of the gyp joint known as the industrial economy.

The first item on the agenda, I suggest, is the remaking of the rural neighborhoods and communities. The decay or loss of these has demonstrated their value; we find, as we try to get along without them, that they are worth something to us—spiritually, socially, and economically. And we hear again the voices out of our cultural tradition telling us that to have community, people don't need a "community center" or "recreational facilities" or any of the rest of the paraphernalia of "community improvement" that is always for sale. Instead, they need to love each other, trust each other, and help each other. That is hard. All of us know that no community is going to do those things easily or perfectly, and yet we know that there is more hope in that difficulty and imperfection than in all the neat instructions for getting big and getting rich that have come out of the universities and the agribusiness corporations in the past fifty years.

Second, the farmers must look to their farms and consider the losses, human and economic, that may be implicit in the way those farms are structured and used. If they do that, many of them will understand how they have been cheated by the industrial orthodoxy of competition—how specialization has thrown them into competition with other farmer-specialists, how bigness of scale has thrown them into competition with neighbors and friends and family, how the consumer economy has thrown them into competition with themselves.

If it is a fact that for any given farm there is a ratio between people and acres that is correct, there are also correct ratios between dependence and independence and between consumption and production. For a farm family, a certain degree of independence is possible and is desirable, but no farmer and no family can be entirely independent. A certain degree of dependence is inescapable; whether or not it is desirable is a question of who is helped by it. If a family removes its dependence from its neighbors—if, indeed, farmers remove their dependence from their families—and give it to the agribusiness corporations (and to moneylenders), the chances are, as we have seen, that the farmers and their families will not be greatly helped. This suggests that dependence on family and neighbors may constitute a very desirable kind of independence.

It is clear, in the same way, that a farm and its family cannot be *only* productive; there must be some degree of consumption. This also is inescapable; whether or not it is desirable depends on the ratio. If the farm consumes too much in relation to what it produces, then the farm family is at the mercy of its suppliers and is exposed to dangers to which it need not be exposed. When, for instance, farmers farm on so large a scale that they cannot sell their labor without enormous consumption of equipment and supplies, then they are vulnerable. I talked to an Ohio farmer recently who cultivated his corn crop with a team of horses. He explained that, when he was plowing his corn, he was *selling* his labor and that of his team (labor fueled by the farm itself and, therefore, very cheap) rather than *buying* herbicides. His point was simply that there is a critical difference between buying and selling and that the name of this difference at the year's end ought to be net gain.

Similarly, when farmers let themselves be persuaded to buy their food instead of grow it, they become consumers instead of producers and lose a considerable income from their farms This is simply to say that there is a domestic economy that is proper to the farming life and that it is different from the domestic economy of the industrial suburbs.

Finally, I want to say that I have not been talking from speculation but from proof. I have had in mind throughout this essay the one example known to me of an American community of small family farmers who have not only survived but thrived during some very difficult years: I mean the Amish. I do not recommend, of course, that all farmers should become Amish, nor do I want to suggest that the Amish are perfect people or that their way of life is perfect. What I want to recommend are some Amish principles:

1. They have preserved their families and communities.
2. They have maintained the practices of neighborhood.
3. They have maintained the domestic arts of kitchen and garden, household and homestead.
4. They have limited their use of technology so as not to displace or alienate available human labor or available free sources of power (the sun, wind, water, and so on).
5. They have limited their farms to a scale that is compatible both with the practice of neighborhood and with the optimum use of low-power technology.
6. By the practices and limits already mentioned, they have limited their costs.
7. They have educated their children to live at home and serve their communities.
8. They esteem farming as both a practical art and a spiritual discipline.

These principles define a world to be lived in by human beings, not a world to be exploited by managers, stockholders, and experts.

Notes

1. In conversation.
2. Robert Heilbroner, "The Act of Work," Occasional Papers of the Council of Scholars (Washington, D.C.: Library of Congress, 1984), p. 20.
3. Eric Gill, *A Holy Tradition of Working* (Suffolk, England: Golgonooza Press, 1983), p. 61.
4. Ibid., p. 65.
5. William Safire, "Make That *Six* Deadly Sins—A Re-examination Shows Greed to Be a Virtue," Courier-Journal (Louisville, Ky.), 7 Jan., 1986.
6. In conversation.
7. Hatch Act, United States Code, Section 361b.
8. Marty Strange, "The Economic Structure of a Sustainable Agriculture," in *Meeting the Expectations of the Land*, ed. Wes Jackson, Wendell Berry, and Bruce Colman (San Francisco: North Point Press, 1984), p. 118.
9. Ibid., p. 116.

15

Conserving Communities*

Wendell Berry

In this chapter, Wendell Berry effectively continues the previous chapter on place-based enterprise by building a set of guidelines for sustainably managing a whole community. Notice that a first step is to attend to the local, land-based economy, especially the food system, recognizing that the ultimate effect is never only local: the health of people and institutions and the maintenance of higher, nonindustrial values are at stake. Echoing Kirkpatrick Sale on decentralism, Berry's reasoning reveals that attending to the local economy—what we might call localizing—is a centuries-old battle against distant, centralized control (see the box titled "From Corn to Veggies: Penalties and Punishments").

In October of 1993, the *New York Times* announced that the United States Census Bureau would "no longer count the number of Americans who live on farms." In explaining the decision, the *Times* provided some figures as troubling as they were unsurprising. Between 1910 and 1920, we had 32 million farmers living on farms—about a third of our population. By 1950, this population had declined, but our farm population was still 23 million. By 1991, the number was only 4.6 million, less than 2 percent of the national population. That is, our farm population had declined by an average of almost half a million people a year for forty-one years. Also, by 1991, 32 percent of our farm managers and 86 percent of our farmworkers did *not* live on the land they farmed.

These figures describe a catastrophe that is now virtually complete. They announce that we no longer have an agricultural class that is, or that can require itself to be, recognized by the government; we no longer have a "farm vote" that is going to be of much concern to politicians.

From Corn to Veggies: Penalties and Punishments

Contrary to what its name might imply, localization is not just about the local. It's about organizing at the local, regional, national, and even international levels. Minnesota farmer Jack Hedin discovered the national dimension of localization when he tried to convert corn acreage to fruit and vegetables for sale at natural food stores and a community-supported agriculture (CSA) program.

The U.S. Agriculture Department's commodity farm program supports, with tax dollars and special rules, the production of corn, soybeans, rice, wheat, and cotton. A farmer under that program who attempts to plant fruits and vegetables has to give up the subsidy. But, in what seems draconian, the farmer is also "penalized the market value of the illicit crop," writes Hedin. The penalties only apply to fruits and vegetables, not to commodity crops and, amazingly, not if the farmer plants nothing at all. For Hedin, a small grower, the penalty one year was $8,771.

To label locally grown and sold fruits and vegetables "illicit" is perplexing, to say the least. But Hedin did some research and feels he has an explanation. "National fruit and vegetable growers based in California, Florida and Texas fear competition from regional producers like myself. Through their control of Congressional delegations from those states, they have been able to virtually monopolize the country's fresh produce markets," he writes.

Hedin and small growers across the country will probably agree that it is not enough to form CSAs and support farmers' markets. Localizers will have to trek to Washington, D.C., to straighten out such antilocalization policies.

Source: Jack Hedin, "My Forbidden Fruits (and Vegetables)," *New York Times*, March 1, 2008.

American farmers, who over the years have wondered whether or not they counted, may now put their minds at rest: they do not count. They have become statistically insignificant.

We must not fail to appreciate that this statistical insignificance of farmers is the successful outcome of a national purpose and a national program. It is the result of great effort and of principles rigorously applied. It has been achieved with the help of expensive advice from university and government experts, by the tireless agitation and exertion of the agribusiness corporations, and by the renowned advantages of competition—of our farmers among themselves and with farmers of other countries. As a result, millions of country people have been liberated from farming, landownership, self-employment, and other idiocies of rural life.

But what has happened to our agricultural communities is not exceptional any more than it is accidental. This is simply the way a large, exploitive, absentee economy works. For example, here is a *New York Times* News Service report on "rape-and-run" logging in Montana:

> Throughout the 1980s, the Champion International Corp. went on a tree-cutting binge in Montana, leveling entire forests at a rate that had not been seen since the cut-and-run logging days of the last century.
>
> Now the hangover has arrived. After liquidating much of its valuable timber in the Big Sky country, Champion is quitting Montana, leaving behind hundreds of unemployed mill workers, towns staggered by despair and more than 1,000 square miles of heavily logged land.

The article goes on to speak of the revival of "a century-old complaint about large, distant corporations exploiting Montana for its natural resources and then leaving after the land is exhausted." And it quotes a Champion spokesman, Tucker Hill, who said: "We are very sympathetic to those people and very sad. But I don't think you can hold a company's feet to the fire for everything they did over the last twenty years."

If you doubt that exhaustion is the calculated result of such economic enterprise, you might consider the example of the mountain counties of eastern Kentucky from which, over the last three-quarters of a century, enormous wealth has been extracted by the coal companies, leaving the land wrecked and the people poor.

The same kind of thing is now happening in banking. In the county next to mine an independent local bank was recently taken over by a large out-of-state bank. Suddenly some of the local farmers and small-business people, who had been borrowing money from that bank for twenty years and whose credit records were good, were refused credit because they did not meet the requirements of a computer in a distant city. Old and once-valued customers now find that they are known by category rather than character. The directors and officers of the large bank clearly have reduced their economic thinking to one very simple question: "Would we rather make one big loan or many small ones?" Or to put it only a little differently: "Would we rather support one larger enterprise or many small ones?" And they have chosen the large over the small.

This economic prejudice against the small has, of course, done immense damage for a long time to small or family-sized businesses in city and country alike. But this prejudice has often overlapped with an industrial prejudice against anything rural and against the land itself, and this prejudice has resulted in damages that are not only extensive but also long-lasting or permanent.

As we all know, we have much to answer for in our use of this continent from the beginning, but in the last half-century we have added to our desecrations of nature a deliberate destruction of our rural communities. The statistics I cited at the beginning are incontrovertible evidence of this. But so is the condition of our farms and forests and rural towns. If you have eyes to see, you can see that there is a limit beyond which machines and chemicals cannot replace people; there is a limit beyond which mechanical or economic efficiency cannot replace care.

I am talking here about the common experience, the common fate, of rural communities in our country for a long time. It has also been, and it will increasingly be, the common fate of rural communities in other countries. The message is plain enough, and we have ignored it for too long: the great, centralized economic entities of our time do not come into rural places in order to improve them by "creating jobs." They come to take as much of value as they can take, as cheaply and as quickly as they can take it. They are interested in "job creation" only so long as the jobs can be done more cheaply by humans than by machines. They are not interested in the good health—economic or natural or human—of any place on this earth. And if you should undertake to appeal or complain to one of these great corporations on behalf of your community, you would discover something most remarkable: you would find that these organizations are organized expressly for the evasion of responsibility. They are structures in which, as my brother says, "the buck never stops." The buck is processed up the hierarchy until finally it is passed to "the shareholders," who characteristically are too widely dispersed, too poorly informed, and too unconcerned to be responsible for anything. The ideal of the modern corporation is to be (in terms of its own advantage) anywhere and (in terms of local accountability) nowhere. The message to country people, in other words, is this: Don't expect favors from your enemies.

And that message has a corollary that is just as plain and just as much ignored: The governmental and educational institutions from which rural people should by right have received help have not helped. Rather than striving to preserve the rural communities and economies and an adequate rural population, these institutions have consistently aided, abetted, and justified the destruction of every part of rural life. They have eagerly served the superstition that all technological innovation is good. They have said repeatedly that the failure of farm families, rural businesses, and rural communities is merely the result of progress and efficiency and is good for everybody.

We are now pretty obviously facing the possibility of a world that the supranational corporations, and the governments and educational systems that serve them, will control entirely for their own enrichment—and, incidentally and inescapably, for the impoverishment of all the rest of us. This will be a world in which the cultures that preserve nature and rural life will simply be disallowed. It will be, as our experience already suggests, a postagricultural world. But as we now begin to see, you cannot have a postagricultural world that is not also postdemocratic, postreligious, postnatural—in other words, it will be posthuman, contrary to the best that we have meant by "humanity."

In their dealings with the countryside and its people, the promoters of the so-called global economy are following a set of principles that can be stated as follows. They believe that a farm or a forest is or ought to be the same as a factory; that care is only minimally necessary in the use of the land; that affection is not necessary at all; that for all practical purposes a machine is as good as a human; that the industrial standards of production, efficiency, and profitability are the only standards that are necessary; that the topsoil is lifeless and inert; that soil biology is safely replaceable by soil chemistry; that the nature or ecology of any given place is irrelevant to the use of it; that there is no value in human community or neighborhood; and that technological innovation will produce only benign results.

These people see nothing odd or difficult about unlimited economic growth or unlimited consumption in a limited world. They believe that knowledge is property and is power, and that it ought to be. They believe that education is job training. They think that the summit of human achievement is a high-paying job that involves no work. Their public boast is that they are making a society in which everybody will be a "winner"—but their private aim has been to reduce radically the number of people who, by the measure of our historical ideals, might be thought successful: the independent, the self-employed, the owners of small businesses or small usable properties, those who work at home.

The argument for joining the new international trade agreements has been that there is going to be a one-world economy, and we must participate or be left behind—though, obviously, the existence of a one-world economy depends on the willingness of all the world to join. The theory is that under the rule of international, supposedly free trade, products will naturally flow from the places where they can be best produced to the places where they are most needed. This theory assumes the long-term safety and sustainability of massive international transport, for which there are no guarantees, just as there are no guarantees that products will be produced in the best way or to the advantage of the workers who produce them or that they will reach or can be afforded by the people who need them.

There are other unanswered questions about the global economy, two of which are paramount: How can any nation or region justify the destruction of a local productive capacity for the sake of foreign trade? and How can people who have demonstrated their inability to run national economies without inflation, usury, unemployment, and ecological devastation now claim that they can do a better job in running a global economy? American agriculture has demonstrated by its own ruination that you cannot solve economic problems just by increasing scale and, moreover, that increasing scale is almost certain to cause other problems—ecological, social, and cultural.

We can't go on too much longer, maybe, without considering the likelihood that we humans are not intelligent enough to work on the scale to which we have been tempted by our technological abilities. Some such recognition is undoubtedly implicit in American conservatives' longstanding objection to a big central government. And so it has been odd to see many of these same conservatives pushing for the establishment of a supranational economy that would inevitably function as a government far bigger and more centralized than any dreamed of before. Long experience has made it clear—as we might say to the liberals—that to be free we must limit the size of government and we must have some sort of home rule. But it is just as clear—as we might say to the conservatives—that it is foolish to complain about big government if we do not do everything we can to support strong local communities and strong community economies.

But in helping us to confront, understand, and oppose the principles of the global economy, the old political alignments have become virtually useless. Communists and capitalists are alike in their contempt for country people, country life, and country places. They have exploited the countryside with equal greed and disregard. They are alike even in their plea that it is right to damage the present in order to make "a better future."

The dialogue of Democrats and Republicans or of liberals and conservatives is likewise useless to us. Neither party is interested in farmers or in farming or in the good care of the land or in the quality of food. Nor are they interested in taking the best care of our forests. The leaders of these parties are equally subservient to the supranational corporations. Of this the North American Free Trade Agreement and the new revisions to the General Agreement on Tariffs and Trade are proof.

Moreover, the old opposition of country and city, which was never useful, is now more useless than ever. It is, in fact, damaging to everybody involved, as is the opposition of producers and consumers. These are not differences but divisions that ought not to exist because they are to a considerable extent artificial. The so-called urban economy had been just as hard on urban communities as it has been on rural ones.

All these conventional affiliations are now meaningless, useful only to those in a position to profit from public bewilderment. A new political scheme of opposed parties, however, is beginning to take form. This is essentially a two-party system, and it divides over the fundamental issue of community. One of these parties holds that community has no value; the other holds that it does. One is the party of the global economy; the other I would call simply the party of local community. The global party is large, though not populous, immensely powerful and wealthy, self-aware, purposeful, and tightly organized. The community party is only now becoming aware of itself; it is widely scattered, highly diverse, small though potentially numerous, weak though latently powerful, and poor though by no means without resources.

We know pretty well the makeup of the party of the global economy, but who are the members of the party of local community? They are people who take a generous and neighborly view of self-preservation; they do not believe that they can survive and flourish by the rule of dog eat dog; they do not believe that they can succeed by defeating or destroying or selling or using up everything but themselves. They doubt that good solutions can be produced by violence. They want to preserve the precious things of nature and of human culture and pass them on to their children. They want the world's fields and forests to be productive; they do not want them to be destroyed for the sake of production. They know you cannot be a democrat (small d) or a conservationist and at the same time a proponent of the supranational corporate economy. They believe—they know from their experience—that the neighborhood, the local community, is the proper place and frame of reference for responsible work. They see that no commonwealth or community of interest can be defined by greed. They know that things connect—that farming, for example, is connected to nature, and food to farming, and health to food—and they want to preserve the connections. They know that a healthy local community cannot be replaced by a market or an entertainment industry or an information highway. They know that contrary to all the unmeaning and unmeant political talk about "job creation," work ought not to be merely a bone thrown to the otherwise unemployed. They know that work ought to be necessary; it ought to be good; it ought to be satisfying and dignifying to the people who do it, and genuinely useful and pleasing to the people for whom it is done.

The party of local community, then, is a real party with a real platform and an agenda of real and doable work. And it has, we might add, a respectable history in the hundreds of efforts, over several decades, to preserve local nature or local health or to sell local products to local consumers. Now such efforts appear to be coming into their own, attracting interest and energy in a way they have not done before. People are seeing more clearly all the time the connections between conservation and economics. They are seeing that a community's health is largely determined by the way it makes its living.

The natural membership of the community party consists of small farmers, ranchers, and market gardeners, worried consumers, owners and employees of small shops, stores, community banks, and other small businesses, self-employed people, religious people, and conservationists. The aims of this party really are only two: the preservation of ecological diversity and integrity, and the renewal, on sound cultural and ecological principles, of local economies and local communities.

So now we must ask how a sustainable local community (which is to say a sustainable local economy) might function. I am going to suggest a set of rules that I think such a community would have to follow. And I hasten to say that I do not consider these rules to be predictions; I am not interested in foretelling the future. If these rules have any validity, it is because they apply now.

If the members of a local community want their community to cohere, to flourish, and to last, these are some things they would do:

1. Always ask of any proposed change or innovation: What will this do to our community? How will this affect our common wealth?

2. Always include local nature—the land, the water, the air, the native creatures—within the membership of the community.

3. Always ask how local needs might be supplied from local sources, including the mutual help of neighbors.

4. Always supply local needs *first*. (And only then think of exporting. First to nearby cities and then to others.)

5. Understand the unsoundness of the industrial doctrine of "labor saving" if that implies poor work, unemployment, or any kind of pollution or contamination.

6. Develop properly scaled value-adding industries for local products to ensure that the community does not become merely a colony of the national or global economy.

7. Develop small-scale industries and businesses to support the local farm and/or forest economy.

8. Strive to produce as much of the community's own energy as possible.

9. Strive to increase earnings (in whatever form) within the community and decrease expenditures outside the community.

10. Make sure that money paid into the local economy circulates within the community for as long as possible before it is paid out.

11. Make the community able to invest in itself by maintaining its properties, keeping itself clean (without dirtying some other place), caring for its old people, teaching its children.

12. See that the old and the young take care of one another. The young must learn from the old, not necessarily and not always in school. There must be no institutionalized "child care" and "homes for the aged." The community knows and remembers itself by the association of old and young.

13. Account for costs now conventionally hidden or "externalized." Whenever possible, these costs must be debited against monetary income.

14. Look into the possible uses of local currency, community-funded loan programs, systems of barter, and the like.

15. Always be aware of the economic value of neighborly acts. In our time the costs of living are greatly increased by the loss of neighborhood, leaving people to face their calamities alone.

16. As a rural community, always be acquainted with, and complexly connected with, community-minded people in nearby towns and cities.

17. Formulate an economy that will always be more cooperative than competitive, for a sustainable rural economy is dependent on urban consumers loyal to local products.

These rules are derived from Western political and religious traditions, from the promptings of ecologists and certain agriculturists, and from common sense. They may seem radical, but only because the modern national and global economies have been formed in almost perfect disregard of community and ecological interests. A community economy is not an economy in which well-placed persons can make a "killing." It is not a killer economy. It is an economy whose aim is generosity and a well-distributed and safeguarded abundance. If it seems unusual to hope and work for such an economy, then we must remember that a willingness to put the community ahead of profit is hardly unprecedented among community businesspeople and local banks.

How might we begin to build a decentralized system of durable local economies? Gradually, I hope. We have had enough of violent or sudden changes imposed by predatory interests outside our communities. In many places, the obvious way to begin the work I am talking about is with the development of a local food economy. Such a start is attractive because it does not have to be big or costly, it requires nobody's permission, and it can ultimately involve everybody. In does not require us to beg for mercy from our exploiters or to look for help where consistently we have failed to find it. By "local food economy" I mean simply an economy in which local consumers buy as much of their food as possible from local producers and in which local producers produce as much as they can for the local market.

Several conditions now favor the growth of local food economies. On the one hand, the costs associated with our present highly centralized food system are going to increase. Growers in the Central Valley of California, for example, can no longer depend on an unlimited supply of cheap water for irrigation. Transportation costs can only go up. Biotechnology, variety patenting, and other agribusiness innovations are intended not to help farmers or consumers but to extend and prolong corporate control of the food economy; they will increase the cost of food, both economically and ecologically.

On the other hand, consumers are increasingly worried about the quality and purity of their food, and so they would like to buy from responsible growers close to home. They would like to know where their food comes from and how it is produced. They are increasingly aware that the larger and more centralized the food economy becomes, the more vulnerable it will be to natural or economic catastrophe, to political or military disruption, and to bad agricultural practice. For all these reasons, and others, we need urgently to develop local food economies wherever they are possible. Local food economies would improve the quality of food. They would increase consumer influence over production; consumers would become participatory members in their own food economy. They would help to ensure a sustainable, dependable supply of food. By reducing some of the costs associated with long supply lines and large corporate suppliers (such as packaging, transportation, and advertising), they would reduce the cost of food at the same time that they would increase income to growers. They would tend to improve farming practices and increase employment in agriculture. They would tend to reduce the size of farms and increase the number of owners.

Of course, no food economy can be, or ought to be, *only* local. But the orientation of agriculture to local needs, local possibilities, and local limits is indispensable to the health of both land and people, and undoubtedly to the health of democratic liberties as well.

For many of the same reasons, we need also to develop local forest economies, of which the aim would be the survival and enduring good health of both our forests and their dependent local communities. We need to preserve the native diversity of our forests as we use them. As in agriculture, we need local, small-scale, nonpolluting industries (sawmills, woodworking shops, and so on) to add value to local forest products, as well as local supporting industries for the local forest economy.

Just as support for sustainable agriculture should come most logically from consumers who consciously wish to keep eating, so support for sustainable forestry might logically come from loggers, mill workers, and other employees of the forest economy who consciously wish to keep working. But *many* people have a direct interest in the good use of our forests: farmers and ranchers with woodlots, all who depend on the good health of forested watersheds, the makers of wood products, conservationists, and others.

What we have before us, if we want our communities to survive, is the building of an adversary economy, a system of local or community economies within, and to protect against, the would-be global economy. To do this, we must somehow learn to reverse the flow of the siphon that has for so long been drawing resources, money, talent, and people out of our countryside with very little if any return, and often with a return only of pollution, impoverishment, and ruin. We must figure out new ways to fund, at affordable rates, the development of healthy local economies. We must find ways to suggest economically—for finally no other suggestion will be effective—that the work, the talents, and the interest of our young people are needed at home.

Our whole society has much to gain from the development of local land-based economies. They would carry us far toward the ecological and cultural ideal of local adaptation. They would encourage the formation of adequate local cultures (and this would be authentic multiculturalism). They would introduce into agriculture and forestry a sort of spontaneous and natural quality control, for neither consumers nor workers would want to see the local economy destroy itself by abusing or exhausting its sources. And they would complete at last the task of gaining freedom from colonial economics, begun by our ancestors more than two hundred years ago.

IV

Bringing Out the Best in People

It is neither wealth nor splendor, but tranquility and occupation which give happiness.

—Thomas Jefferson (July 12, 1788)

To understand the usefulness of the next three readings, it is necessary to consider two prevailing perspectives on human nature—the individualistic and the "new consciousness." The individualistic says that humans are egocentric, short-term gain maximizers. As such, they have evolved to consume resources with little concern for waste, always trying to pass costs on to others and regularly forming exclusive groups that neglect the interests of outsiders. According to this perspective, attending to long-term societal or environmental benefit is simply not in the equation for humans.

That humans can and do act this way is without question. But when discussing human behavior, statements that say "our species' behavior is to always do X" should be considered suspect. The brain is more malleable, and behavior more adaptive, than such narrow statements allow. What's more, the error of this pessimistic perspective is compounded by the advice that typically follows: when problems arise, people's behavior must be manipulated; they must be managed using incentives, disincentives, tight prescriptive rules, and moralistic norms, or the content of their mental models (e.g., attitudes, values, worldviews) must be altered. Alas, even if the assumed privilege of the would-be manipulator is granted, research shows that such manipulations are not reliable and rarely durable.

The second prevailing perspective—the "new consciousness"—seems more hopeful. It begins with the prospect of a revolutionary change in human consciousness. It presumes that humans can, and soon will, evolve to interact in entirely new ways. They will experience the environment in a holistic manner, exercise heroic self-restraint, and achieve a full understanding of their place in the universe.

This optimistic perspective does have historical precedents—the ethical developments over the centuries underlying universal human rights and the abolition of slavery, for instance. Two things, however, are troublesome. First, the evolutionary transformation is always presumed to occur sometime in the future. To date, only ambiguous instances of such fundamental change have occurred. Second, a novel element of human cognition is presumed, one that is not yet known but would be essential—for example, a new form of awareness, a new motive, a new value. Enough is known from research about the human psyche to cast grave doubt on any solution to the world's problems that starts with the need for something novel but as yet undiscovered in humans' psychological makeup. To plan as if such an element will emerge just when needed is foolhardy, especially when faced with environmental dilemmas as serious as climate disruption and resource depletion.

The readings in this section take a different stance. They are based on well-established features of human behavior (e.g., prevailing values, existing inclinations, universal motivations). One theme across all three readings is that useful motives for bringing out the best in people are well known and can be found at the core of many everyday behaviors. Psychologists refer to them as *intrinsic satisfactions*. They exist and they are universal; they do not have to be invented or forced (see box titled Work Less, Consume Less, Live Better).

Work Less, Consume Less, Live Better

Simple living advocates have argued that people in Northern industrial work-and-spend societies could all cut back and still live better. If the premises of this book bear out (i.e., that there will be an unavoidable downshift in consumption), then we will soon find out. But we don't have to wait for the future. A glance at the past, in particular at the life of a company town, suggests what is possible—namely below-average work hours, below-average consumption, and above-average quality of life.

Battle Creek, Michigan, has been the home of Kellogg Company's ready-to-eat breakfast cereal for nearly a century. When, in the 1920s and 1930s, the country, especially business, was debating what to do about excess production capacity and insufficient consumer demand (i.e., excess frugality), W. K. Kellogg and the company president Lewis Brown decided to reduce work hours rather than lay off workers. They wanted to show that the "free exchange of goods, services, and labor in the free market would not have to mean mindless consumerism or eternal exploitation of people and natural resources," as author Benjamin Hunnicutt put it [in his book, *Kellogg's Six-Hour Day*]. Rather, "workers would be liberated by increasingly higher wages and shorter hours for the final freedom promised by the Declaration of Independence—the pursuit of happiness." Brown wrote to his employees about the "mental income" of "the enjoyment of the surroundings of your home, the place you work, our neighbors, the other pleasures you have [that are] harder to translate into dollars and cents." All this would lead to "higher standards in school and civic . . . life."

Kellogg's bold experiment cut across the grain of American business, especially the newly emerging fields of retail and marketing aimed at spurring consumption and, it might have seemed, work hours. But Kellogg

Work Less, Consume Less, Live Better
(continued)

succeeded—for the company, the workers and the broader community. One business reporter found "a lot of gardening and community beautification, athletics and hobbies . . . libraries well patronized and the mental background of these fortunate workers . . . becoming richer." Canning became popular for many, not just to preserve food but, as Hunnicutt writes, as a "medium for . . . [sharing] stories, jokes . . . practical instruction, songs, griefs, and problems." A U.S. Department of Labor study of the Kellogg Company found "little dissatisfaction with lower earnings resulting from the decrease in hours." And after long work hours during World War II, in a vote by workers in 1946, 77 percent of men and 87 percent of women wanted to return to a thirty-hour week.

Writer Jeffrey Kaplan sums up Kellogg's unusual practices: "This was the stuff of a human ecology in which thousands of small, almost invisible, interactions between family members, friends, and neighbors create an intricate structure that supports social life in much the same way as topsoil supports our biological existence. When we allow either one to become impoverished, whether out of greed or intemperance, we put our long-term survival at risk."

According to an overwhelming body of scientific evidence, both our social and biological existence is indeed being impoverished by working too many hours and consuming too many products. Maybe the cereal town in Michigan offers a better way.

Source: Jeffrey Kaplan, "The Gospel of Consumption: And the Better Future We Left Behind," *Orion*, May/June 2008.

The difficulty for localizers is that these intrinsic satisfactions are internal; their voices are often quiet and easily overlooked in high-powered business and policy environments or in the glitter of modern commerce and entertainment. Intrinsic satisfactions are easily displaced by events and signals coming from a hostile external environment, including employment settings where self-promotion is rewarded. And yet, as Thomas Jefferson and many others have known for a long time, deep contentment derives from simple life patterns, ones that secure the well-being of self, family, and community. These observations are indeed hopeful signs for localization. But do notice that they are not based on top-down mandates, the enlightened prescriptions of a benevolent authority, external rewards and punishments, the cleverness of a technological elite, or the optimism of those who anticipate a new stage in human awareness. Rather, they are based on what drives ordinary people to work hard and work together—the desire to thrive.

If we editors were to offer one piece of advice to localizers based on the common perspective of these readings, it would go something like this: Do not downplay the physical and social challenges of sustainable living. Instead, support peoples' innate desire to understand, explore, and find meaning in everyday activities. Validate the quiet, internal voice that finds satisfaction, even fascination, in life patterns that are demanding and consequential.

16

Abundance and Fulfillment*

Sharon Astyk

Localization is not about the pursuit of pleasure, but something much deeper. Writer and farmer Sharon Astyk begins this chapter with an emotion few people in the global change community wish to confront—fear. She admits to fear when confronted by peak oil and climate disruption. Nevertheless, she explains, as a parent, she cannot allow fear to paralyze her. Rather, she acknowledges the motivation that a sense of duty conveys and takes a strong moral position.

Astyk feels that the great abundance and opportunity Americans now have create a great obligation to repair the world. They must take a lead role in the coming transition to durable living. She is not looking to the elites for direction, but to ordinary people who can be galvanized by such an effort. And while the task of repairing the world will be hard work, the meaningfulness of that work can provide an antidote to fear as well as a deep and enduring sense of satisfaction at day's end.

Scared? Duh!

I was at a conference recently, and I did something that conference speakers aren't supposed to do. I admitted that I was afraid. Conference speakers are supposed to be inspiring, and admitting your fear is high on the disheartening meter. But it is true, and I doubt I'm the only one.

The thing is, I think most people have a choice when confronted by a reality like Peak Oil and Climate Change—either they develop a thick skin for at least some things, or they deny. We're all aware that denial is

*Sharon Astyk, "Abundance, democracy, joy." In *Depletion and abundance: Life on the new home front,* 231–241. Gabriola Island: New Society Publishers, 2008. Excerpted and reprinted by permission.

the most popular choice, and why not—denial is a very happy place to live, as long as you don't mind the cost. But there is a high cost—we can't begin to mend the damage we've done from a position of denial.

I'm an optimist by nature, and I've also got a lot of practice laughing rather than weeping. Or I wind up my computer and type out my outrage hoping to break through someone else's denial and make them as angry as this stuff makes me, so that maybe we can do something to stop it. But I'm less practiced at dealing with my own fear. I've been dripping with outrage at the world's injustices pretty much since I was old enough to have a political conscience, sometime in my early teens. But I haven't been particularly scared, because my own blood was never in the game. I was always outraged on someone else's behalf, and of course, that's an easy emotion.

I have, as we all know, four little hostages to fortune and at least the average person's fear of suffering, death and inconvenience. I'm scared for my kids and scared for myself. Some of the time I desperately wish this would all just go away and I wouldn't have to think about it anymore. Sometimes I wish denial were an option.

But mostly, I'm glad I know even the bad, hard stuff. Because I honestly have no doubt that being prepared is better than not being prepared. I'm not even always sure I want more time. Part of me does, but part of me believes we are better off going through our depletion crisis sooner than later—soon enough that we still have money and resources to make some major infrastructure changes, soon enough that we may avoid the worst of catastrophic warming, and that there might be enough oil left for future generations for some wind power and vaccinations. And I can't wish my knowledge would go away because it is my job to protect my kids, and my desire to protect the next generations in general—I don't want to dump this burden on my sons or on other people's children. I don't think that's the proper work of parents who love their kids.

John Adams once said that he was a soldier so that his son could be a farmer, and his grandson a poet. I'm no soldier, and if this were war (it isn't) it would be won by farmers and perhaps by poets too. But I share the sentiment. I'm going to do this work and face this as head on as I possibly can so that my children may someday choose other work. That's what moms do. Now, the thing they don't tell you about parenting when you become a mom or dad is this: being a parent doesn't make you a better person. . . .

I know the world is full of better people than I am, but the truth is that a lot of us are still the same ordinarily rotten people we were before we had kids. We just don't have the option of indulging our rottenness. That is, parenthood requires not that you be a good person or that your better nature predominate, but that you suck it up and do the unselfish thing anyway, even when you don't want to, even when it is damned hard.

The same is true about our present situation. We've got bad news, and it is appropriate to feel bad about it. There's no reason we have to be fearless here—frankly, the only way I can imagine being fearless is to be stupid. But we do have to be brave—that is, we don't have to feel brave but, like the Cowardly Lion, like the mom who doesn't really want to get up for the two a.m. feeding, we have to act the right way, to pretend as hard as we can that we have, as the Cowardly Lion's song says, the nerve. And the amazing thing about pretending hard is that sometimes—not always, but just sometimes—you become, as Kurt Vonnegut put it, "what you pretend to be."

The only antidote to fear I know is good work. I learned in pregnancy, facing labor (all of my labors were very, very, very long), to simply screw up my nerve, accept that the only way out is through, and to go forward into the pain. We're in the same situation now—the way out of this current crisis is through it, to go forward from where we are, with what we have and who we are. It isn't required that we not be afraid or that we don't spend a lot of time grumpily wishing that someone else would do the work and leave us alone with our book. But it is required that while we curse fate, previous generations, the current administration, G-d and the Federal Reserve, we get to work.

What work? Tikkun olam, if you are a Jew, or even if you find the metaphor compelling. *Tikkun olam* means "the repair of the world." In my faith, that is why we are here—to fix what is broken, repair what is damaged, to improve what can be improved. As the saying goes, it is not required of us that we complete the work, but it is not permitted for us not to try.

I do not come from one of those religious faiths where you put aside the lesser emotions like fear and selfishness—in fact, as far as I can tell, the right to whine is a sacrament in Judaism. So I'd hardly be the person to tell anyone "Don't be afraid." Instead, I suggest we all be afraid. Nor do I suggest any of us fail to whine about it—that, after all, is what the Internet and best friends are there for.

But let us whine while we hammer, moan while we cook, sigh in outrage while we write and march and yell and build and fight our fear with good work and the pretense that maybe we'll become better people while we're pretending that we already are. There's too damned much to do to do it any other way. I may be a coward at times (and trust me, I am), but I've got work to do anyway. . . .

Abundance

Peter Parker said . . . "with great power comes great responsibility." I think most of us have no idea how powerful we are, and thus, how responsible we are. Virtually all Americans command power and wealth unimaginable to most of the people in the world. We have, as James Kunstler has pointed out, the equivalent of 200 slaves working for us—but instead of human slaves, we have energy slaves that wash our clothes, wash our dishes, make the clothes, carry us about. Most of us have more education—even if we graduated only from high school—than a majority of the world's population. It doesn't feel that way when you are in debt and struggling economically, but most middle-class Americans are richer and more privileged than kings in most of history.

Because we do not see ourselves as powerful and rich (we view ourselves mostly in comparison to our neighbors who are similarly powerful and rich), we are all caught up in our struggles; we do not tend to think that we are the very people who have great responsibility in the world. Other people are powerful. Other people can change things, not us. We are merely getting along, we do not have time, we do not have energy, we do not have money enough to spare.

But if we do not, who on earth has the time and the money, the energy and the power to change the world? Who will you ask to do it for you? Will you ask someone poorer and weaker and less privileged? In many cases those people are already doing this work—all over the world, the poor have spoken up about Climate Change and world trade, land reform and sustainability.

I have read analyses of global warming and the WTO written by 12-year-olds from Nicaragua and India that put the writing of professional adults to shame.

The world is full of people who work harder than you and I, who have harder lives, fewer electronic slaves, . . . yet who still have time to stand up and speak out.

I have written this elsewhere, but I repeat it, and will keep repeating it as long as necessary: almost all that is good in human history over the past three centuries has been accomplished by oppressed and frightened, impoverished and angry people who have stood up to those that did them harm, who mortgaged their futures and endangered their lives and said "No More." Overwhelmingly, they succeeded in winning, despite a lack of things you and I have plentifully—power, money, education, comfort. Our own national history includes, along with its dark side, a remarkable and courageous tradition of not counting the cost to do what is right. And every single person who has ever stood up in resistance has been less well educated, less wealthy, less privileged, less safe, less comfortable than you and I are today. How can we do less?

Most of us are not living up to our moral responsibilities or using our privilege and wealth to create justice. . . . We are afraid of change, afraid of doing without, afraid of being different. The thought that we might have to give up all the things we are accustomed to and change to something entirely new is frightening. So mostly we are silent.

. . . Whether you believe in God or good fortune, the randomness of everything or some sort of intentionality, perhaps if we are very lucky, it is because our good fortune enables us to bring about change. Perhaps we are meant to lead, no matter how little we like the work, how frightened we are of the consequences or how comfortably we are ensconced in the dominant culture.

. . . We are afraid of what it would mean to reveal ourselves, to stand forth from the culture and demand that it change. . . . But . . . sometimes what happens to us isn't really the point—sometimes what matters is that we, in our power, have done the right thing, without counting the cost to ourselves. It takes courage. And that is not in over-great supply. But I suspect there is more of it out there than we like to admit, even to ourselves.

Am I Romanticizing Poverty?

Someone who reads my blog recently e-mailed me that my writings are merely a call for us all to return to poverty, and that I'm intentionally romanticizing subsistence agriculture. And I started wondering, am I?

The answer, I suspect, is a little bit, in the sense that I don't think anything is served by my saying, "Your future and the future of your children is drudgery and misery." I think it is certainly possible that I elide some difficulties—or rather, that I prefer not to focus on them. Some of that is the optimist in me. And part of it is that ultimately most of the things that will necessarily get harder aren't the things I value most. That is, I suspect our physical loads will get heavier. On the other hand, I suspect that will be good for my overall health and wellbeing, so I choose to look at it as mostly a positive.

Some things about a life low on the economic food chain, I think, really are better. For example, poor agrarian societies generally have stronger social ties. In many cases, people who live in simpler economies, enticed with fewer things they can't have, report themselves to be happier. And the things about contemporary, wealthy society that really matter are mostly things that we can continue to have—if we are very careful. The things that wealth has given us that I value are these: basic medical care, including birth control and preventive care; social support networks for the elderly, the disabled, the very poor and other vulnerable people; good education; access to information; access to clean water; safe food and secure shelter; personal freedom; and a just society. And what is fascinating about all these things is that they aren't very expensive. A good education, up to and including college, doesn't have to cost 30K a year. Basic public medical care including vaccinations, preventive medicine, midwifery, simple palliative care for the dying, many basic medications, birth control and some hospital care doesn't have to cost us what it does. Neither do libraries, public services and support programs for the poor.

Most of the most important things in my life are items that are not depleted or in short supply. As Richard Heinberg has put it,

> Are there some good things that are not at or near their historic peaks? I can think of a few:
>
> - Community
> - Personal autonomy
> - Satisfaction from honest work well done
> - Intergenerational solidarity
> - Cooperation
> - Leisure time
> - Happiness
> - Ingenuity
> - Artistry
> - Beauty of the built environment (Heinberg, 2007, 14)

The blunt truth is that an abundance of the things above is enough to compensate for the loss of other gifts.

It is worth remembering that when the Soviet Union collapsed and stopped supplying oil to Cuba, crashing the economy and everything along with it, the Cuban government did exactly the opposite of what the American government does in hard times—it kept up the social support programs. Instead of taking much-needed funds out of education, social welfare, programs for the elderly and poor, it kept those up. It opened new university campuses and more clinics because people couldn't travel as far or as easily for medical care and education. That's a choice we can make too—if we want to.

On the other hand, am I going to deny that our wealth has been extremely pleasant? Heck no. I've enjoyed all sorts of things other people can never imagine. I've traveled. I've had pretty things. I have a home of my own. I have a computer to write on and the Internet. Right now I'm sitting here on a 15-degree day, two sleeping dogs at my feet, in a warm house typing and listening to The Little Willies. Would I prefer to be outside, hand pumping icy water into buckets and carrying it?

If that were my life, if I were hauling water in the cold instead of writing here, would I be unhappy? Maybe momentarily, but generally speaking, I don't think so. I like personal comfort as much as anyone else, but, as trite as it seems to say so, the things I really care about don't depend on my not having to grow food or haul water.

Our perceptions drive our sense of what is work more than the actual work does. How many people can remember doing some now-unthinkable job when they were young and poor, and now say, "But I was happy." I've met people who walked in the snow to their outhouses, who boiled laundry on coal stoves, who hung their dripping, freezing laundry off a fourth-story balcony. And I've hauled a month's worth of laundry half a mile on my back in a sack, carried my groceries for a mile, stood outside in the cold waiting for a bus every morning, walked four miles to work. And when I look back at every one of those activities, it really wasn't that big a deal.

We look back on what we used to do and think "Amazing. We were happy. All that work didn't impinge on our enjoyment of life." But what's amazing is what we've forgotten—that work really doesn't impinge on enjoying life when it is our life. We take on our labor savers as though they are miracles, but the life we had before them was usually not so very bad. The miracle, if you can call it that, is that they've reshaped our memories so that our pasts are untenable, and untenantable, to us—we begin to think we can't go home again.

Do I romanticize subsistence agriculture? Maybe a little. I like farming, and someone who doesn't might not agree with me. And I tend to think that if we're going to have to do something (and I have little doubt that we will have to), we might as well go into it excited, treating it as an opportunity to optimize and improve our lives, rather than as a tragedy to be endured.

But I also note that I'm happier since we moved here. And I think this might not be a purely personal preference. Some of you may have watched the PBS documentary series *Frontier House*. Like all such things, it was imperfect in its creation, to some degree more about the personalities than the work. It was originally intended to debunk the myth of *Little House on the Prairie*, offering counterweight to the romanticism of subsistence agriculture. And in the end, it failed to do so—in fact, it proved that that romanticism wasn't entirely misplaced.

At the end of six months without any of the amenities of 21st-century life, without indoor plumbing or refrigeration, thermostats or grocery stores, seven adults and six children came out of the experience changed. A majority of the adults and all the children overwhelmingly found that they preferred their frontier lives to the ones they returned to. One of the men actually moved back to live in his old cabin and help out on the ranch where the filming had occurred. Another child experienced serious depression because she missed the life she described as more "real." Overwhelmingly, the kids on the show said that they missed having chores, they missed taking care of animals, and they missed being with their parents all the time. (This included multiple teenagers.) A wealthy woman building a 5,000-square-foot house admitted that her house felt too big, and that in a 400-square-foot cabin, six people had never felt crowded.

Now, *Frontier House* was television, but what matters about it is how thoroughly it failed to do what it set out to do. The producers had assumed that the physical hardships would overwhelm every other part of the experience. They did for a short adjustment period, and then the emotional, spiritual and personal benefits of the life overtook the transitory concerns of physical work, and again, life was good.

So maybe I'm a little romantic. But I draw hope that if we may not be more comfortable, we might still be having fun.

The One Thing We Did Right

I've been watching the rerelease of *Eyes on the Prize*, which I haven't seen since high school. I recommend to everyone that you watch it too.

Not only is it a brilliant representation of our history and one of the best documentaries of all time, it is also an inspiration for the future.

Peak Oil is not about petroleum geology or economics when you get right down to it. Climate Change is not about ice cores and meteorology. Those things matter, but they aren't the center of things. Peak Oil and Climate Change are about justice, plain and simple. They are about fairness, morality and integrity—we in the rich world have chosen to steal from the poor in our own country and other nations, and from our children and grandchildren, and we need to stop it right now.

The stakes are very simple: our children's lives, other people's lives, the food in their mouths and the medicine that keeps them from dying unnecessarily. If we keep consuming resources as though there is no tomorrow, there will be no tomorrow, and those who are too young or too weak or too powerless to demand anything be saved for them will die. They are dying right now, today, in poor world countries as we in the west extract $38 billion of wealth from them every year.

And that is a drop in the bucket compared to the number who stand to suffer and die because of Climate Change, Peak Oil and economic disruption. It can't always be someone else. It will be my kids and yours. We have to make deep changes, and we have to make them now.

If you believe we can't find the strength and courage and commitment to give up our cars, our heat or air conditioning, or our jobs that produce nothing and give wealth to the corporations we pretend to deplore, go watch *Eyes on the Prize* right now. Watch a 65-year-old woman with diabetes and varicose veins tell with pride how she walked eight miles round trip to her job scrubbing floors every day for more than a year, and never, ever took a ride on a bus no matter how tired she was. Watch a seven-year-old girl walk past a row of people screaming obscenities at her and throwing things, just to go to school. Watch an old man face death threats to walk into a courtroom to testify to the truth. See people face dogs and firehoses and men with guns who want to kill them and link arms and march forward. We all know people did this, because we read about it in our history classes, but what this documentary does better than any other single source is show how ordinary those actions are.

Those people were no different than you or me under the skin. They were ordinary people with ordinary fears and an extraordinary degree of courage. And I do not believe for one moment that those remarkable people, or the remarkable young men who faced death in World War II, or any of the heroes of history are any different than you and I. We too can have courage. We too can have justice. We too can do what has to be done.

What would the world look like if all of us who worry about Peak Oil and Climate Change showed true integrity? What would it look like if the millions of people who know what is coming refused to go on warming the planet and burning fossil fuels, and pledged to find another way? What would happen if we had the courage of our convictions and stood up and said "I will no longer steal from the future and the poor. I will live only on what is mine by right and in justice." There is no doubt in the world we could do this, because people like us already have. I'm trying to find out how to get there. I hope you will too.

References

Heinberg, Richard. 2007. *Peak Everything: Waking Up to the Century of Declines*. Gabriola Island, British Columbia, Canada: New Society Publishers.

17

Motives for Living Lightly*

Raymond De Young

Earlier in this book, Rob Hopkins (chapter 5) proposed several energy-descent scenarios. Some pointed to chaos while others led to positive outcomes. To be helpful, positive scenarios must be not only technically feasible, but psychologically viable.

This chapter explores two scenarios, both from the perspective of psychological well-being. One involves ecological living, the other a pro-technology outlook. Notice that sustainable living, far from requiring great sacrifice, contains its own reward in the form of intrinsic satisfaction. Notice also that the fundamental motive at work here is not altruism, but indeed the inverse, a form of self-interest in which the behaviors are pursued because they create direct, personal, and internal contentment. Localizers may find this good news because, during an energy and resource descent, material and other tangible rewards may be scarce. Nonmaterial, intrinsically satisfying rewards, on the other hand, will be abundant.

With survival having always depended on the careful stewardship of finite resources, one might expect people to have come to recognize the sorts of life patterns where such care was both possible and supported. But it is important for people to not only recognize such patterns; they should also find them satisfying to pursue.

The study of the relationship between lifestyles and environmental stewardship behavior has received attention. It has been noted, for instance, that lifestyles have a significant effect on energy consumption over and above that explained by income and energy pricing (Schipper

*Adapted from De Young, Raymond. 1990/1991. "Some psychological aspects of living lightly: Desired lifestyle patterns and conservation behavior." *Journal of Environmental Systems* 20: 215–227. Excerpted and reprinted with permission.

et al. 1989). Some of this attention was inspired by the work of Gregg on voluntary simplicity (Gregg 1974a, 1974b). The core of a deliberately simple life is argued to be frugality—the intentional avoidance of wasteful practices. Frugality itself received considerable attention during the 1960s as people began to question the appropriateness of a high-consumption and high-waste lifestyle (Henderson 1978, Inglehart 1977). Frugality has more recently been characterized as a central aspect of a sustainable society (Henion and Kinnear 1979) as well as a goal worthy of national attention (Johnson 1978, 1985).

While frugality may be accepted as a necessary feature of sustainable living, it is usually portrayed as an onerous undertaking, one requiring personal sacrifice of the highest order. People, it is argued, are being asked to give up a modern, high-technology existence for an austere, bleak but needed substitute. And it is here that the greatest resistance to the widespread adoption of sustainable living is thought to exist. It is imagined that people will adopt such a pattern of behavior *only* if they receive just compensation. Seligman has captured the essence of this conserve-only-if-compensated argument:

> Unless business can make money from environmental products or politicians can get elected on environmental issues, or *individuals can get personal satisfaction from experiencing environmental concern*, then individuals and organizations will simply do whatever competes with environmentalism if they see the pay off as greater (cited in Geller 1990). [emphasis added]

Deeply embedded here is the presumption of a direct link between our individual consumption behavior and well-being. For some, the reasons why people would both willingly and without tangible compensation adopt a simple, frugal lifestyle remains a mystery. Yet an answer to this mystery is no more complicated than that found in the emphasized portion of the Seligman quote above. As many people throughout time have suggested, and Gregg has written, a rich inner sense of well-being can be gained from a simple approach to life (Gregg 1974a, 1974b).

Thus the issue becomes not one of compensation, but of whether a frugal pattern of living, acknowledged as being environmentally appropriate, might also provide for a sense of personal satisfaction. A number of studies have explored this question. De Young (1996, 2000) has reviewed a set of these studies and discusses the research methods used in the investigation reported on below. This investigation explored what future life patterns were more preferred and the relationships between these patterns and intrinsic satisfactions.

What Was Measured

The survey instrument contained groups of items that were designed to measure the constructs of desired lifestyle patterns and intrinsic satisfactions and that were drawn from earlier research on conservation behavior (De Young 1986, 1985–1986). All items used a 5-point Likert rating scale, adjusted during analysis so that a score of 5 always indicates high endorsement for a construct. The groups of items (e.g., desired patterns, satisfactions) were subjected to dimensional analysis; robust categories were identified and are reported in the tables below.

The construct of desired patterns dealt with how technology should develop in the future and with what patterns of person-environment interaction would be most preferred. While this desired pattern construct may appear on the surface to be attitudinal in nature, it is more appropriate to consider it something akin to the personal goals or purposes an individual holds. Attitudes are commonly characterized as being evaluative in nature, as involving a favorable or unfavorable reaction to an object, event or behavior. Broader purposes and goals, on the other hand, can be thought of as reflecting a state of affairs one would like to experience, a particular way of life or a system of priorities (Leiss 1980–1981). To capture the sense of future orientation this construct embodies, the survey respondents were asked to imagine the future as they would like it to be and then to indicate to what extent it would include each of the listed items. Included were items which dealt with desire for pastoralism, self-reliance, acceptance of ecological limits, faster travel and communication modes, etc. In addition the respondents were asked how they would like to see technology developed and used.

The satisfaction items covered the personal satisfaction derived from avoiding waste, keeping things working long past their normal life, doing things which help bring order to the world, having the luxuries and conveniences of our society, being a member of an affluent society, etc.

The respondents were 56 percent women and roughly 80 percent were homeowners. Most were long-time residents (more than 47 percent said they had lived in town for more than 20 years). Some 16 percent of the sample were under thirty years old, 50 percent were in their thirties or forties, 16 percent were in their fifties, and 18 percent were sixty or older.

The Desired Pattern Categories

The analysis identified four distinct categories named Ecological Lifestyle, Pro-Technology, Self-Reliance and Waste-is-OK (see table 17.1). The Ecological Lifestyle category may be thought of as measuring a desire to develop a sustainable human-environment relationship. The Pro-Technology category gauges the degree to which techniques and products of the industrial age will continue to underpin our well-being. Self-Reliance captures a desire to depend upon no one other than oneself. Waste-is-OK embodies a desire to consume freely provided none suffer.

The difference in category mean values for every pair-wise comparison is significant at $p < 0.02$, suggesting that the respondents have a higher preference for an ecologically-appropriate way of life.

The Intrinsic Satisfaction Categories

Three satisfaction categories emerged from the survey data (table 17.2). Satisfaction from *Frugality*, defined as the prudent use of resources, is closely tied to everyday living, involving such things as what items one buys, what activities one pursues, and what one does with waste materials. Furthermore, the chance for *Participation*, to be involved and have their actions contribute to the preservation of the environment, is viewed positively by the respondents. . . . The respondents also reported deriving satisfaction from *Luxuries*. It is interesting to note that the satisfaction from *Luxuries* category has very low correlations with each of the other satisfaction categories. This supports the notion that satisfaction derived from luxuries may not always be in direct opposition to satisfaction gained from sustainable living.

Relationships between Desired Patterns and Satisfaction

The desired pattern categories are further understood by examining the relationships they have with the satisfaction categories. As shown in table 17.3, those respondents who most longed for an ecologically-focused or self-reliant pattern of living reported the highest scores on the satisfaction from *Frugality* and *Participation* categories. The *Pro-Technology* and *Waste-is-OK* categories related to the satisfaction categories in a manner quite different from the relationship exhibited by the other desired pattern categories. Those respondents with higher scores on the

Table 17.1
Desired Pattern Categories

Category name and items included	Mean	S.D.	Alpha*
ECOLOGICAL LIFESTYLE	4.20	.65	.85
Things would last longer			
Would avoid creating pollution			
Cooperate with nature			
More time to reflect on things			
Repair and maintain things, not always start anew			
Less messiness in alleyways, etc.			
Spend more time getting to know others			
Conservation would be part of our culture			
Care more about function than image			
PRO-TECHNOLOGY	3.51	.76	.85
Technology will solve resource scarcity problems			
Machines separate and sort our waste			
Use technology to explore space			
Increase our manufacturing efficiency			
Build longer lasting goods			
Increase our standard of living			
Extract hard-to-get resources			
Explore new fields, ideas, etc.			
Have machines do our manual labor			
Use technology to keep us alive longer			
SELF-RELIANCE	2.92	.98	.79
Save more of our income			
People would make their clothing			
Grow more of our own food			
People would make many of the things they need			
WASTE-IS-OK	2.06	1.06	.60
OK to waste money if earned it			
OK to use more than need if none suffer			

*Cronbach's coefficient alpha, a measure of the degree of coherence within the category.

Table 17.2
Satisfaction Categories

Category name and items included	Mean	S.D.	Alpha
Intrinsic satisfaction derived from:			
FRUGALITY	3.78	.81	.87
Keeping things running past normal life			
Finding ways to avoid waste			
Repairing rather than throw away			
Saving things I might need someday			
Doing things which don't rely on others			
Finding ways to use things over and over			
PARTICIPATION	3.65	.88	.93
Reduce pressure on Earth to supply needs			
Helping make sense out of our world			
Fitting into our place in natural scheme			
Taking actions that can change our world			
Do things that help bring order to world			
Not pushing resource scarcity onto future			
Influencing how society solves problems			
Reducing dependency on scarce resources			
Doing things that matter in the long run			
Living by an ecological ethic			
LUXURIES	2.97	.78	.83
Having clothing that is in style			
Having new items to try, evaluate, and buy			
Having vast resources at our disposal			
Having many choices when buying			
Having luxury/conveniences of our society			
Using latest electronic consumer product			
Knowing we are looked upon as affluent			

Pro-Technology and *Waste-is-OK* categories tended to have significantly higher scores on the satisfaction from *Luxuries* category.

Conclusion

The desired pattern categories reflect the system of priorities the respondents said they preferred. The most preferred is an ecological lifestyle, a pattern which scored higher than a technology-based lifestyle. This expressed preference may be of particular interest to policymakers

Table 17.3
Means on Satisfaction as Function of Response to Desired Pattern*

Desired Pattern Categories		Satisfaction Categories		
		Frugality	Participation	Luxury
Ecological Lifestyle	High	4.2	4.1	
	Med	3.9	3.8	
	Low	3.3	3.0	
Pro-Technology	High			3.4
	Med			2.9
	Low			2.5
Self-Reliance	High	4.1	4.1	
	Med	3.8	3.5	
	Low	3.4	3.3	
Waste-is-OK	High			3.1
	Low			2.8 **

*All relationships are significant at $p < .0001$ unless noted. Blank cells are not significant at $p < .05$
** Significant at $p < .01$

given the fact that a high technology existence would seem to best characterize the immediate future. It turns out that people have longings for particular patterns of interaction with the environment which may not be adequately addressed by many development policies and plans. And rather than being of interest to active conservers alone, such a high level of concern for ecological issues has been reported for conservers and non-conservers alike (De Young 1988–1989, Vining and Ebreo 1990). One might be justified in assuming that this is a dominant value.

The pattern of self-reliance—including elements of sustainable living that were much discussed in the early days of the environmental movement, and sometimes tied to a back-to-the-land perspective—was significantly less preferred than the ecological pattern. Together, these data suggest that technological advancement and self-sufficiency, so much a part of the American lore, may not be as central an issue as is the concern for human-environment compatibility. This is not to suggest that technology has no role to play in the respondents' lives. The *Ecological Lifestyle* was not highly correlated with the *Pro-Technology* category, positively or negatively. This suggests that the respondents do not view an ecologically compatible existence as necessarily devoid of all technology, only that technology is not a key feature of such an existence.

Taken together, these findings provide an interesting contrast to the dominant view of resource conservation. In analyses of conservation behavior, research often groups conservation strategies into two categories: behavior that reduces the use of resources, referred to as curtailment, and behavior that involves the adoption of advanced resource-efficient technologies, considered efficiency strategies. Stern and Gardner (1981) state that "when people decrease their use of existing energy systems, they see themselves as making do with less—curtailing the benefits derived from energy use; when they adopt more efficient technologies, they are getting more benefits from the same energy expenditure or the same benefits for less energy." And it has been suggested that people will never willingly adopt curtailment strategies.

The distinction between curtailment and efficient use is sometimes useful when designing behavior change strategies. However, as Winett and Geller (1981) have pointed out, a word such as curtailment has powerful, often negative, connotations (e.g., sacrifice, freezing in the dark). This is an unfortunate way to view the conservation behavior that is at the core of sustainable living. It reinforces the worn adage that reducing one's resource consumption can only cause a reduction in one's quality of life and sense of well-being.

The findings reported here address this issue directly and offer a distinctly different perspective. Rather than equating conservation with sacrifice and hence demanding compensation for such extraordinary behavior, the respondents associate forms of intrinsic satisfaction with a reduced consumption pattern of living. Thus, perhaps, the dichotomy of curtailment versus efficient use might sometimes prove unnecessary as we promote sustainable living.

As the magnitude of the environmental dilemmas we face is realized, there are increasingly vigorous efforts to promote environmental stewardship behavior. This promotion has tended to focus upon the necessity of such activities, with efforts made to point out one's ecological duty. It is intriguing to consider an alternative approach. Perhaps one could build upon the possibility that the low consumption society we must create for ourselves is not without its bright points. As Johnson suggests, far from being a great sacrifice, living lightly on the earth may increase our quality of life and sense of well-being (Johnson 1978, 1985).

References

De Young, Raymond. 1985–1986. "Encouraging environmentally appropriate behavior: The role of intrinsic motivation." *Journal of Environmental Systems* 15:281–292.

De Young, Raymond. 1986. "Some psychological aspects of recycling: The structure of conservation satisfactions." *Environment and Behavior* 18:435–449.

De Young, Raymond. 1988–1989. "Exploring the difference between recyclers and non-recyclers: The role of information." *Journal of Environmental Systems*. 18:341–351.

De Young, Raymond. 1996. "Some psychological aspects of reduced consumption behavior." *Environment and Behavior* 28:358–409.

De Young, Raymond. 2000. "Expanding and evaluating motives for environmentally responsible behavior." *Journal of Social Issues* 56:509–526.

Geller, Scott E. 1990. "Behavior analysis and environmental protection: Where have all the flowers gone?" *Journal of Applied Behavior Analysis* 23: 269–273.

Gregg, Richard. 1974a. "Voluntary simplicity: Part I." *MANAS*, September 4.

Gregg, Richard. 1974b. "Voluntary simplicity: Part II." *MANAS*, September 11.

Henderson, Carter. 1978. "The frugality phenomenon." *Bulletin of the Atomic Scientists* 34:24–27.

Henion, Karl E., and Thomas Kinnear, eds. 1979. *The Conserver Society*. Chicago, IL: American Marketing Association.

Inglehart, Ronald. 1977. *The Silent Revolution*. Princeton, NJ: Princeton University Press.

Johnson, Warren. 1978. *Muddling Toward Frugality*. Boulder, CO: Shambhala.

Johnson, Warren. 1985. *The Future Is Not What It Used to Be: Returning to Traditional Values in an Age of Scarcity*. New York: Dodd, Mead and Company.

Leiss, William. 1980–1981. "A value basis for conservation policy." *Journal of Policy Studies* 9:613–620.

Schipper, Lee, Sarita Bartlett, Dianne Hawk, and Edward Vine. 1989. "Linking life-styles and energy use: A matter of time?" *Annual Review of Energy* 14: 273–320.

Stern, Paul C., and Gerald T. Gardner. 1981. "Psychological research and energy policy." *American Psychologist* 36:329–342.

Vining, Joanna, and Angela Ebreo. 1990. "What makes a recycler?: A comparison of recyclers and nonrecyclers." *Environment and Behavior* 22:55–73.

Winett, Richard A., and E. Scott Geller. 1981. "Comment on 'Psychological research and energy policy.'" *American Psychologist* 36:425–426.

18

Enabling the Best in People*

Rachel Kaplan and Stephen Kaplan

Unreasonable behavior (e.g., being irresponsible, unhelpful, intolerant, domineering) seems to be rampant in fast-paced, high-consuming industrialized societies. One might conclude that it is humans' standard operating condition. Many years of psychological research, however, show this conclusion to be wrong. Environmental psychologists Rachel Kaplan and Stephen Kaplan suggest that the difference between reasonable and frustratingly unreasonable behavior may be partly explained by the environments in which people find themselves. This leads the Kaplans to identify the salient features of environments that support reasonable behavior. Notice that their insights have great potential for localization when the term environment *is taken in a general sense, not just in the sense of nature.*

The Kaplans's model of human nature is based on inherited inclinations: (1) humans seek meaningfulness more than enjoyment and (2) they benefit more from a sense of competence, clarity, and mental vitality than from convenience or hedonic pleasure. The mind is better adapted to exploring, problem solving, and sense making than it is to affluence. Yet, in consumerist, growth-, and convenience-oriented societies, these universal inclinations are easily neglected, if not diminished, as familiar as they otherwise may be.

In such societies, the coming transition will dramatically alter the context and content of everyday behavior. Surprising to some, these societal changes can actually stimulate people's natural inclinations to explore and understand, to achieve clarity, and to pursue acts of meaning. Thus, the transition will create many of the very conditions that, research

*Kaplan, Rachel, and Steven Kaplan. 2008. "Bringing Out the Best in People: A Psychological Perspective," *Conservation Biology* 22: 826–829. Excerpted and reprinted with permission.

shows, support reasonable behavior. The challenge for localizers is to recognize and promote these positive inclinations, knowing that unreasonable behavior can also emerge.

David Orr's (2008) call for a greater role for psychology in efforts to ensure "a decent future" for humanity is right on. Psychology has much to offer to the understanding of and approaches to survival and sustainability . . .

There are many ways to consider psychology's potential contribution. Although Orr casts these in terms of the study of "mind," we have found it more effective to talk about "human functioning" and specifically to consider what brings out the best in people. Many of Orr's examples point to the opposite—situations (e.g., Milgram study) that have brought out undesirable qualities in people. Although situations are important influences, it is by no means the case that their outcomes are uniformly (or even predominantly) negative. Rather, there is a wide range of situations that can bring out the best in people. This is a particularly critical issue when one thinks about endangered environments. Natural environments, for example, have been repeatedly shown to have the capacity to bring out the best in people even when nature is no more than the view of a tree from a window (Kaplan 1993, Frumkin 2001, Kaplan and Kaplan 2005).

As for an emphasis on human selfishness and self-interest, this too provides only a partial view of our species. It is true that advertising and the media have adeptly orchestrated a materialistic culture in which consumption has become an escapist distraction relative to the values that are essential for a sustainable world. Here, perhaps, neoclassical economics helps explain behavior. The behavioral economists, however, have identified a broad range of circumstances in which people do not maximize their personal gain, but rather are cooperative and reasonable (Hammerstein and Hagen 2005). For example, research on how people deal with antisocial behavior shows that punishing violators is experienced as pleasurable even when it costs in terms of one's material standing (Angier 2002). The vast body of research on human well-being has identified many dimensions of this complex domain, including hedonic pleasure, emotional well-being, psychological capital, and character strengths (e.g., Ring et al. 2007, Kesebir and Diener 2008, Ryff and Singer 2008). Particularly pertinent to the discussion here is that many human qualities are inconsistently exhibited within the same person across time and circumstances. In other words, the dimensions of well-being or of what people find satisfying are not immutable; rather, they are strongly influenced by context or situation (Luthans et al. 2007).

From our perspective this is central to the issue of what brings out the best in people. In our experience psychology can provide a better understanding of the contexts, or environments, that are more likely to achieve survival and sustainability. Such outcomes, however, are not necessarily or primarily concerned with individual well-being. Maximizing individual well-being can, and often has, come at a great cost; vast global inequities are an iconic expression. Rather, we see that a more benign and more effective route to well-being involves maximizing the larger social good. The agenda must be to foster reasonableness (e.g., being responsible, cooperative, and tolerant). There is reason to believe that a great deal of human satisfaction and well-being flows from participation in actions that improve the lives of others.

The reasonable person model (RPM) is a framework we have been developing for understanding these bringing-out-the-best-in-people contexts (Kaplan and Kaplan 2003, R. and S. Kaplan, unpublished data). These contexts are ones that help meet basic human needs, and the needs are conceptualized in terms of one of the most pervasive characteristics of the species—concern about and dependence on information. Information is what we store, trade, hide, and act on. We are overwhelmed by it, yet endlessly seek it. We cannot act without it.

The RPM organizes human informational needs into three major categories. These are highly interconnected and often interdependent, yet they address different domains that are all important for fostering reasonableness. These needs relate to understanding what is going on around one (*building models*), the capacity to utilize knowledge and skill (*being effective*), and the desire to be needed and to make a difference (*meaningful action*).

Building Models

By models we mean mental models, the basic structures of the mind that store knowledge and its affective qualities. Craik (1943, 61) summarizes the essential qualities of these models eloquently:

If the organism carries a "small scale model" of external reality and of its own possible actions within its head, it is able to try out various alternatives, conclude which is the best of them, react to future situations before they arise, utilize the knowledge of past events in dealing with the present and future, and in every way to react in a much fuller, easier, and more competent manner to the emergencies which face it.

Building these models is a lifelong enterprise. The psychology of survival is intimately related to appreciation of the complexities of this building program. There are no kits for these models or easy mechanisms for transporting them from one head to another. A vast body of psychological research tells us that the mental models develop with experience. In other words, not only is understanding achieved with repeated exposure to aspects of the model's domain, but exploration of the domain is more effective than others' efforts to deliver the model (Bransford et al. 2000).

In our view, at the core of the public's inertia about daunting global events is the lack of adequate mental models. People are constantly bombarded with conflicting information about climate change, biodiversity, and other major environmental issues. Much of the information is provided with technical details and little imagery about ways to explore the pertinent domains. People dislike feeling confused and overwhelmed; when information is not meaningful, it is likely to be ignored.

Similarly, efforts to persuade people to change their attitudes are also often unlikely to be fruitful because the messages fail to connect with the mental models of the intended recipients. There is considerable evidence that people deeply resent being told what to do (Brehm 1966); it is not only teenagers who may decide under such circumstances to do the opposite of what they are told! Thus, a more appropriate alternative is to view the process as one of helping a fellow human build a model. This model must make sense relative to the experience the individual already has. This suggests the importance of listening (an activity that is readily neglected among those eager to insert their ideas and beliefs in the minds of others). The desired attitude change is also more likely to take shape if the recipient is given a chance to explore options and to take small steps in the desired direction.

Being Effective

People also dislike feeling incompetent. Being overwhelmed by information, and especially information that is confusing, seemingly urgent, and largely uninterpretable, is unlikely to contribute to clearheadedness or a sense of competence. It is hardly surprising that people in such situations can become unpleasant, uncooperative, and even destructive. Being civil is difficult when the mind is noisy with confusion and despair. Thus, being an information-based species comes with the necessity to cope with managing information.

It is useful to distinguish between two dimensions of information in understanding how to manage our dependence on it while also coping with its (over)abundance. The distinction is derived from the kinds of attention needed to process information. Much of the time the information surrounding people requires their attention; one needs to listen to a lecture or an endless conversation, to read something intently to use it in a document, or to simultaneously attend to different demands. Such information takes focus and effort and is referred to as directed attention. A person can maintain or direct attention for only so long before becoming less effective (i.e., before attention capacity declines). People recognize this condition as being tired or stressed out. Interestingly, however, fatigue is differential. People are not too tired to resume reading a novel, to solve a crossword puzzle, or to go on a walk and take pictures of trees in bloom. Such tasks also require processing information, but the information is inherently fascinating and thus is effortless.

Attention restoration theory (ART; Kaplan and Kaplan 1989/1995, Kaplan 1995) contends that attention plays a key role in why mental fatigue occurs and in how restorative environments can foster recovery. Mental fatigue is the result of declining capacity to use directed attention. Recovering from it calls on the other kind of attention, attention to activities that are fascinating and compelling. Attention restoration theory posits that time spent in such effortless pursuits and contexts is an important factor in the recovery from mental fatigue. In other words, restoration involves activities and settings that are compelling and allow directed attention to rest. Tending to our attentional needs is central for achieving clearheadedness. There is a substantial empirical literature, drawing on ART, that documents the important role the natural environment can play in recovering attentional capacities (Tennessen and Cimprich 1995, Kuo and Sullivan 2001, Cimprich and Ronis 2003, Berto 2005).

Meaningful Action

News about catastrophes and calamities, war and terror, and abuse and corruption foster a sense of hopelessness and helplessness. In their despair about chaos and tragedy, people often reach out to try to salvage something small and graspable. There are many examples of when substantial amounts of money are contributed to help an individual who survived a disaster when hundreds of others may have perished. People have a need to find meaning and to act in ways that enhance that meaningfulness. The acts may be small or big, lasting a moment (e.g., voting) or lasting a long time. They may involve people one knows well or strangers; they may be directed to other people or to many other outcomes. For example, volunteers involved in environmental stewardship believe they are improving the environment and contributing to a better world (Grese et al. 2000). Such actions give one a sense that one can make a difference, that one participates in a larger whole.

Hopefulness abounds in the cornucopia of examples and stories of people's actions to address the array of contemporary woes. Bornstein's (2004) book offers stunning examples of individuals who have made a difference in ways that have multiplied many times over as their efforts have found replication. Hawken's (2007) *Blessed Unrest* bears a subtitle that speaks directly to these issues: *How the Largest Movement in the World Came into Being and Why No One Saw It Coming*. That movement is the groundswell of meaningful action, the thousands of organizations that have sprung up around the world to address aspects of these woes. The Web site behind this effort (www.wiserearth.org) invites viewers to explore "108,705 organizations, 12,621 people, 526 groups, 353 events, and 143 jobs." The areas of focus are organized under several dozen major headings (e.g., agriculture and farming, biodiversity, coastal and marine ecosystems, community development, and energy) with numerous subheadings for each. It is an impressive anthology of positive examples and of the human inclination to act meaningfully.

We know of no empirical evidence, but one would suspect that those who are deeply involved in these organizations and activities are likely to feel less helpless about the state of the world. Still, even when one has experience with what to do, it may not be sufficient to counter the sinking feeling that it cannot possibly be enough, that no matter the size of the effort, the negative forces are too strong and coming too fast to be stopped by individuals' actions. But what if the media have been falling down on the job? What if they have been failing to tell us how many exciting, ingenious, and well-financed projects are already under way and that their potential is huge? This is precisely the message of Krupp and Horn's (2008) book, *Earth, the Sequel*. They provide an eminently readable survey of the multiplicity of innovative projects involving six different forms of energy. Although the frustration the book evokes concerning the failure of the media is warranted, the hopefulness it fosters leads one to believe that the cumulative effects of individual efforts may well matter. Krupp and Horn provide a basis for believing that the whole is far greater than the sum of the parts. Together, if everyone does their part, humanity and the ecosystems it depends on may have a good chance of survival after all.

Although discussed in the context of meaningful action, the examples we provide here also illustrate the vital roles of model building and being effective. The three domains of RPM are interdependent. Active involvement in an organization calls on exploration and fosters greater understanding, which can enhance one's sense of competence and inspire sustained engagement in an activity.

References

Angier, N. 2002. "Why we're so nice: we're wired to cooperate." *New York Times*, July.

Berto, Rita. 2005. "Exposure to restorative environments helps restore attentional capacity." *Journal of Environmental Psychology* 25(3): 249.

Bornstein, David. 2004. *How to Change the World: Social Entrepreneurs and the Power of New Ideas*. New York: Oxford University Press.

Bransford, John D., Ann L. Brown, and Rodney R. Cocking. 2000. *How People Learn: Brain, Mind, Experience, and School*. Washington D.C.: National Academy Press.

Brehm, Jack W. 1966. *A Theory of Psychological Reactance*. New York: Academic Press.

Cimprich, Bernadine, and David L. Ronis. 2003. "An environmental intervention to restore attention in women with newly diagnosed breast cancer." *Cancer Nursing* 16:83–92.

Craik, Kenneth James Williams. 1943. *The Nature of Explanation*. London: Cambridge University Press.

Frumkin, Howard. 2001. "Beyond toxicity: human health and the natural environment." *American Journal of Preventive Medicine* 20:234–242.

Grese, Robert. E., Rachel Kaplan, Robert L. Ryan, and Jane Buxton. 2000. "Psychological benefits of volunteering in stewardship programs." In *Restoring Nature: Perspectives from the Social Sciences and Humanities*, ed. Paul H. Gobster and R. Bruce Hull, 265–280. Washington D.C.: Island Press.

Hammerstein, Peter, and Edward H. Hagen. 2005. "The second wave of evolutionary economics in biology." *Trends in Ecology & Evolution* 20:604–609.

Hawken, Paul. 2007. *Blessed Unrest: How the Largest Movement in the World Came into Being and Why No One Saw It Coming*. New York: Viking.

Kaplan, R. 1993. "The role of nature in the context of the workplace." *Landscape and Urban Planning* 26:193–201.

Kaplan, Rachel, and Stephen Kaplan. 1989/1995. *The Experience of Nature: A Psychological Perspective.* New York: Cambridge University Press. Republished by Ulrich's, Ann Arbor, Michigan.

Kaplan, Rachel, and Stephen Kaplan. 2005. "Preference, restoration, and meaningful action in the context of nearby nature." In *Urban Place: Reconnecting with the Natural World*, ed. Perry F. Barlett, 271–298. Cambridge: MIT Press.

Kaplan, Stephen. 1995. "The restorative benefits of nature: toward an integrative framework." *Journal of Environmental Psychology* 15:169–182.

Kaplan, Stephen, and Rachel Kaplan. 2003. "Health, supportive environments, and the reasonable person model." *American Journal of Public Health* 93: 1484–1489.

Kesebir, Pelin, and Ed Diener. 2008. "In pursuit of happiness: empirical answers to philosophical questions." *Perspectives on Psychological Science* 3:117–125.

Krupp, Fred, and Miriam Horn. 2008. *Earth, The Sequel: The Race To Reinvent Energy and Stop Global Warming.* New York: Norton.

Kuo, Frances E., and William C. Sullivan. 2001. "Aggression and violence in the inner city: impacts of environment via mental fatigue." *Environment and Behavior* 33:53–57.

Luthans, Fred, Bruce J. Avolio, James B. Avey, and Steven M. Norman. 2007. "Positive psychological capital: measurement and relationship with performance and satisfaction." *Personnel Psychology* 60:541–572.

Orr, David W. 2008. "The psychology of survival." *Conservation Biology* 22: 819–822.

Ring, Lena, Stefan Hofer, Hanna McGee, Anne Hickey, and Ciaran O'Boyle. 2007. "Individual quality of life: can it be accounted for by psychological or subjective well-being?" *Social Indicators Research* 82:443–461.

Ryff, Carol D., and Burton H. Singer. 2008. "Know thyself and become what you are: a eudaimonic approach to psychological well-being." *Journal of Happiness Studies* 9:13–39.

Tennessen, Carolyn M., and Bernadine Cimprich. 1995. "Views to nature: effects on attention." *Journal of Environmental Psychology* 15:77–85.

V

Appropriate Governance

Previous readings have alluded to issues of self-organization and self-rule. One might read them to suggest that, because localization is inevitable in some form, appropriate forms of governance will spontaneously emerge. We don't think so. Positive localization—that is, a peaceful, democratic, just, and ecologically resilient transition to less material and energy use—will emerge *if* people consciously and conscientiously build the needed governing structures, and do so *before* they are compelled to by circumstances. If the task is left to the centralizers, those devoted to continuing business as usual for as long as possible, effective localized governance and positive localization are unlikely to develop. To be sure, localizers will have to seek support from others—neighbors and governments (local, national, international)—but we expect they will have to manage the transition mostly by themselves. And they will have to do so in ways that are appropriate to their own particular tasks, just as technologies and businesses must be appropriate to their tasks. And the overriding task under the premises of this book is to craft a durable prosperity within biophysical limits.

To this end, here we offer several perspectives on the challenge of governance, especially self-governance in a localizing world. Notice that no author claims such governance will be easy. In fact, they all see a lot of messy, complex, tedious negotiating among a range of stakeholders. Also notice that the intended outcome is not stasis, not a final, unchanging end state. Self-governance, with its many networked connections to other institutions at all levels, will require constant attention, adjustment, and negotiation. It will be, like the ecosystems in which they are embedded, dynamic.

Finally, notice that none of the authors are talking about local government per se—how to better run a city, for instance. In fact, they put their ideas in the context of national and international affairs, pointing not at more enlightened central power (e.g., the greening of councils, legislatures, agencies, and transnational corporations), but at the necessity of locally tailored governance. Scale, once again, is crucial: not just the scale of technologies and consumption rates, but the scale of the units and means of governance.

19

Ecological Democracy*

John S. Dryzek

Political theorist John S. Dryzek presents a form of governance called "discursive democracy" and argues that it is better than alternative governance models at solving problems involving natural systems. Thus, this form of democracy may prove particularly useful as communities respond to emerging biophysical limits. Although Dryzek does not address localization directly, his model of democracy would seem to function well in decentralized settings where people and nature will need to interact more frequently.

Notice that discursive democracy requires effective integration of political and ecological communication where listening and negative, self-correcting feedback are central. Barriers to such communication are large-scale organization and remote interaction, creating inattention to the social and biophysical contexts of behavior—what Dryzek refers to as market autism. Appropriate scale is thus crucial to both authentic democracy and sustainable practice.

. . . Inasmuch as there is a conventional wisdom on the matter of ecology and democracy, it would draw a sharp distinction between procedure and substance. As Robert Goodin (1992, 168) puts it, "To advocate democracy is to advocate procedures, to advocate environmentalism is to advocate substantive outcomes." The more general case that in a democracy we cannot pre-specify any particular outcome is made by Robert Dahl (1989, 191). Thus there can never be any guarantee that democratic procedure will produce ecologically benign substance. This distinction between procedure and substance forms the core of Goodin's

*Dryzek, John S. 2001. "Green democracy." In *Deliberative Democracy and Beyond: Liberals, Critics, Contestations,* 140–161. New York: Oxford University Press. Excerpted and reprinted with permission.

(1992b) treatment of green political theory. To Goodin, the green theory of value represents a coherent set of ends related to the protection and preservation of nature. This interpretation of green ends is somewhat narrow. . . . For Goodin, the green theory of agency addresses where and how green values might be promoted. He argues that a green theory of agency cannot be derived from the green theory of value. Greens may still want to advocate (say) grassroots participatory democracy; but they should recognize that any such advocacy has to be on grounds separate from basic green values.

This procedure/substance divide arises most graphically in the context of green advocacy of decentralization and community self-control. Such decentralization of political authority would have decidedly anti-ecological substantive consequences in a lot of places with natural-resource-based local economies. Many counties in the Western United States have tried to assert their authority against federal environmental legislation (so far with little success in the courts) in order that mining, grazing on federal lands, and forest clearcutting can proceed unchecked. This is part of the agenda of the ill-named "Wise Use Movement." Decentralization will only work to the extent local recipients of authority subscribe to ecological values, or, alternatively, the degree to which they must stay put and depend for their livelihoods solely on what can be produced locally.

On this kind of account, political structure obviously matters far less than the adoption of green values on the part of denizens in that structure, or the occupancy of key positions (such as membership in parliament) in that structure by greens. . . .

To begin with the currently dominant order of capitalist democracy, all liberal democracies currently operate in the context of a capitalist market system. Any state operating in the context of such a system is highly constrained in terms of the kinds of policies it can pursue. . . . Policies that damage business profitability—or are even perceived as likely to damage that profitability—are automatically punished by the recoil of the market. Disinvestment here means economic downturn. And such downturn is bad for governments because it both reduces the tax revenue for the schemes those governments want to pursue (such as environmental restoration), and reduces the popularity of the government in the eyes of the voters. This effect is not a matter of conspiracy or direct corporate influence on government; it happens automatically, irrespective of anyone's intentions.

The constraints upon governments here are intensified by the increasing mobility of capital across national boundaries. So, for example, antipollution regulation in the United States stimulates an exodus of polluting industry across the Rio Grande to Mexico's maquiladora sector. Thus irrespective of the ideology of government—and irrespective of the number of green lobbyists, coalition members, or parliamentarians—the first task of any liberal democratic state must always be to secure and maintain profitable conditions for business.

Environmental policy is possible in such states, but only if its damage to business profitability is marginal, or if it can be shown to be good for business. Along these lines, the idea of ecological modernization has recently gained ground in several European states, notably Germany and the Netherlands (Hajer, 1995; Weale 1992, 66–92). United States Vice-President Albert Gore (1992) once pointed to the degree to which environmental protection can actually enhance business profitability. But if ecological modernization is to move beyond isolated successes on the part of green capitalists, it requires a wholesale reorientation of state structure (Christoff, 1996). If green demands are "all or nothing" in Goodin's terms, then "nothing" remains the likely consequence in any clash with economic imperatives.

Even setting aside the economic context of policy determination under capitalist democracy, there remain reasons why the structure of liberal democracy itself is ultimately incapable of responding effectively to ecological problems. To cut a long story short, these problems often feature high degrees of complexity and uncertainty, and substantial collective action problems. Thus any adequate political mechanism for dealing with them must incorporate negative feedback (the ability to generate corrective movement when a natural system's equilibrium is disturbed), co-ordination across different problems (so that solving a problem in one place does not simply create greater problems elsewhere), co-ordination across actors (to supply public goods or prevent the tragedy of the commons), robustness (an ability to perform well across different conditions and contexts), flexibility (an ability to adjust internal structure in response to changing conditions), and resilience (an ability to correct for severe disequilibrium, or environmental crisis) (for greater detail on these requirements, see Dryzek, 1987).

One can debate the degree to which these criteria are met by different political-economic mechanisms, such as markets, administrative hierarchies, and international negotiations, as well as liberal democracies. My own judgment is that liberal democracy does not perform particularly well across these criteria, even when it is organized along the relatively open "passively inclusive" lines of pluralism. . . . Negative feedback under pluralism or polyarchy is mostly achieved as a result of particular actors whose interests are aggrieved giving political vent to their annoyance, be it in voting for green candidates, lobbying, contributing money to the environmental interest groups, or demonstrating. But such feedback devices are typically dominated by the representation of economic interests, businesses and (perhaps) labour. Co-ordination is often problematical because the currency of interest groups pluralism consists of tangible rewards to particular interests. Such particular interests do not add up to the general ecological interest. Further, complex problems are generally disaggregated on the basis of these same particular interests, and piecemeal responses crafted in each of the remaining subsets. The ensuing "partisan mutual adjustment," to use Lindblom's (1965) term, may produce a politically rational resultant. But there is no reason to expect this resultant to be ecologically rational. In other words, interests may be placated in proportion to their material political influence, and compromises may be achieved across them, but wholesale ecological destruction can still result. Resilience in liberal democracy more generally is inhibited by short time horizons (resulting from electoral cycles) and a general addiction to the "political solvent" of economic growth (politics is much happier, and choices easier, when the size of the available financial "pie" is growing).

Despite the ecological inadequacies associated with the interest group pluralism form of liberal democracy, I would argue that among the political mechanisms that have been tried by nation-states from time to time, this form does better than most of the alternatives (see Dryzek, 1987). Only the corporatism associated with ecological modernization may do better (for evidence, see Janicke, Weidner, and Jorgens, 1997). But even setting aside the issue of the ecological adequacy of liberal democracy, and its relative merits compared to other systems, the fact remains that the way political systems are structured can make an enormous difference when it comes to the likelihood or otherwise of realizing green values. And if this is true, then (to use Goodwin's distinction) we should be able to derive an account of the politics from the green theory of value, not just the green theory of agency. Let me now attempt such a derivation, which I will link to an extended discursive democracy. . . .

Just about every human political ideology and political-economic system has at one time or another been justified as consistent with nature, especially nature as revealed by Darwinism.

But this sheer variety should suggest that in nature we will find no single blueprint for human politics. . . .

Yet nature is not devoid of political lessons. What we will find in nature, or at least in our interactions with it, is a variety of levels and kinds of communication to which we humans might try to adapt. The key here is to downplay "centrism" of any kind, and focus instead on the kinds of interactions that might occur across the boundaries between humanity and nature. In this spirit, the search for green democracy can indeed involve looking for progressively less anthropocentric political forms. For democracy can exist not only among humans, but also in human dealings with the natural world—though not *in* that natural world, or in any simple *model* which nature provides for humanity. So the key here is seeking more egalitarian interchange at the human/natural boundary, an interchange that involves progressively less in the way of human autism. In short, ecological democratization here is a matter of more effective integration of political and ecological communication. . . .

The Communicative Rationality of Ecological Democracy

Let me suggest that rather than jettison democracy in the search for an ecologically rational political economy, we might better proceed by detaching democracy from liberal anthropocentrism, while retaining an emphasis on deliberation and communication. . . . A defensible account of deliberative democracy has to be underwritten by a conception of communicative rationality. . . .

Minimally, a recognition of agency in nature would underwrite respect for natural objects and ecological processes. Democrats in general, and deliberative democrats in particular, should of course condemn humans who would silence other humans. Silencing in the form of not allowing others to speak is not the issue when it comes to dealing with communication from the non-human world. Nature "speaks" or does not "speak" irrespective of any attempted human suppression of that ability—indeed, it is not clear what suppression would consist of here, short of destroying nature entirely. But, as Bickford (1996) reminds us, the most effective and insidious way to silence others in politics is a refusal to listen, which is why the practice of effective listening has to be central to any discursive democracy. Recognition of agency in nature therefore means that we should listen to signals emanating from the natural world with the same sort of respect we accord communication emanating from human subjects, and as requiring equally careful interpretation. In other words, our relation to the natural world should not be one of instrumental intervention and observation of results oriented to control. Thus communicative interaction with the natural world can and should be an eminently rational affair (Dryzek, 1990).

Now, it might be argued here that agency as I have defined it is simply the capacity to act instrumentally, in pursuit of some goal, rather than communicatively. However, closer examination shows that communicative capacity too is at issue, even though it is of course not the kind of linguistic communication in which the theory of communicative action is normally grounded. Of course, human verbal communication cannot extend into the natural world. But greater continuity is evident in nonverbal communication—body language, facial displays, pheromones, and so forth (Dryzek 1990, 207). And a lot goes on in human conversation beyond the words, which is why a telephone discussion is not the same as a face-to-face meeting. More important than such continuities here are the ecological processes which transcend the boundaries of species, such as the creation, modification, or destruction of niches; or cycles involving oxygen, nitrogen, carbon, and water. Disruptions in such processes occasionally capture our attention, in the form of (say) climate change, desertification, deforestation, and species extinction. . . .

Democracy Across the Boundary with Nature

. . . In discursive democracy, we look for the essence of democracy not in the aggregation of interests or preferences of a well-defined and well-bounded group of people (such as a nation-state), but rather in the content and style of interactions. Some styles may be judged anti-democratic (for example, the imposition of a decision without possibility for debate or criticism), some relatively democratic (for example, wide dissemination of information about an issue, the holding of hearings open to any interested parties, etc.).

[Elsewhere] I argued against critics who claim that deliberation privileges rational argument by showing that deliberation can accommodate other kinds of voices. A similar extension may be in order to accommodate non-human communication (which Goodin 1996, 841, allows). This extension means that we are now well-placed (or at least better-placed than aggregative liberal democrats) to think about dismantling what is perhaps the biggest political boundary of them all: that between the human and the non-human world. This is indeed a big step, and no doubt some people would still believe that it takes us out of the realm of politics and democracy altogether, at least as those terms are conventionally defined. Yet there is a sense in which human relationships to nature are already political. As Val Plumwood (1995) points out, politicization is a concomitant of human colonization of nature. Such colonization connotes an authoritarian politics; democratization would imply a more egalitarian politics here. . . .

. . . A capacity for effective and egalitarian listening is an essential component of discursive democracy; it is also helpful in undermining unequal power distributions. When it comes to improvement of our social listening capacity in an environmental context a number of institutional devices are already at hand—such as mandatory state-of-the-environment reporting, and cumulative regional impact assessment (Eckersley, 2000).

At one level, it is possible to propose green democracy as a regulative ideal. This is, after all, how the basic principles of both liberal and deliberative democracy can be advanced (Miller 1992, 55–56). For aggregative liberals, the regulative ideal is fairness and efficiency in preference aggregation: the various institutional forms under which reference aggregation might proceed are then a matter for investigation, comparison, and debate. Similarly, for deliberative democrats the regulative ideal is free discourse about issues and interests; again, various institutional forms might then be scrutinized in the light of this ideal. For green democrats, the regulative ideal is effectiveness in communication that transcends the boundary of the human world. As it enters human systems, then, obviously ecological communication needs to be interpreted, and so we move from a politics of presence to a politics of ideas. However, unlike the situation in aggregative liberal democracy, this communication does not have to be mediated by the material interests of particular actors.

The content of such communication might involve attention to feedback signals emanating from natural systems, in which case, the practical challenge when it comes to institutional design becomes one of dismantling barriers to such communication. It is also important to attend to the feedback signal emanating from those closest to environmental damage. Plumwood (1998, 579) calls this the capacity to hear "the bad news from below." With these principles in mind, it is a straightforward matter to criticize institutions that try to subordinate nature on a large scale, and those that are remote and so incapable of hearing any news from below, be it good or bad. Think, for example, of the development projects sponsored by the World Bank, which until recently did not even pretend to take local environmental factors into account (now they at least pretend to), and which have been widely criticized for the social and ecological devastation left in their wake. Yet it is also possible to criticize approaches to our dealings with the environment that do exactly the reverse, and seek only the removal of human agency. On one of his own interpretations, Lovelock's Gaia can do quite well without people. And a misanthrope such as David Ehrenfeld (1978) would prefer to rely on natural processes left well alone by humans.

With this regulative ideal of green democracy in mind, we are, then, in a position both to criticize existing political-economic arrangements and to think about what might work better. The construction of democracy should itself be discursive, democratic, sensitive to ecological signals—and open-ended. Idealist political prescription insensitive to real-world constraints and possibilities for innovation is often of limited value. Further, variation in the social and natural contexts within which political systems operate means that we should be open to institutional experimentation and variety across these contexts (though, as I noted earlier, an ability to operate in different contexts may itself be a highly desirable quality for any political-economic mechanism).

When it comes to criticism of existing political (and economic) mechanisms, it is reasonably easy to use the ecological communicative ideal to expose some gross failings. Perhaps most obviously, to the degree any such mechanism allows internal communications to dominate and distort signals from the outside, then it merits condemnation. So, for example, a bureaucracy with a well-developed internal culture may prove highly inattentive to its environment. And bureaucratic hierarchy pretty much ensures distortion and loss of information across the levels of hierarchy. Indeed, these are standard criticisms of bureaucracy as a problem-solving device, though such criticisms are usually couched in terms of a human environment, not a natural one. Markets can be just as autistic, if in different ways. Obviously, they respond only to human, consumer preferences that can be couched in monetary terms. Any market actor trying to take non-pecuniary actors into account is going to have its profitability, and so survival chances, damaged (this is not to gainsay the possibility of green consumerism). Conversely, the positive feedback of business growth (and the growth of the capitalist market in general) is guided by processes entirely internal to markets.

Above all, existing mechanisms merit condemnation to the extent their size and scope do not match the size and scope of ecosystems and/or ecological problems. Under such circumstances, communications from or about particular ecological problems or disequilibria will be swamped by communications from other parts of the world. Here, markets that transcend ecological boundaries, which they increasingly do, merit special condemnation. The internationalization and globalization of markets makes it that much easier to engage in local despoliation. It may be quite obvious that a local ecosystem is being degraded and destroyed, but "international competitiveness" is a good stick with which to beat environmentalist critics of an operation. For example, they can be told that old growth forests must be clearcut, rather than logged selectively. Obviously, some ecological problems are global, as are some markets. This does not of course mean that effective response mechanisms to global ecological problems can be found in global markets. Market autism guarantees that they cannot.

Turning to the desirable scope and shape of institutions suggested by the ideal of ecological democracy, the watchword here is "appropriate scale." In other words, the size and scope of institutions should match the size and scope of problems. There may be good reasons for the predispositions toward small scale in ecoanarchism and "small is beautiful" green political thought. Most notably, feedback processes in natural systems are diffuse and internal (Patten and Odum, 1981), and do not pass through any central control point. Highly centralized human collective choice mechanisms are not well placed to attend to such diffuse feedback. Moreover, the autonomy and self-sufficiency advocated by green decentralizers can force improved perception of the natural world. To the degree a community must rely on local ecological resources, it will have to take care of them. It does not follow that local self-reliance be taken to an extreme of autarchy. Rather, it is a matter of degree: the more the community is politically and economically self-reliant, then the more it must take care of its local ecosystems. Presumably the degree of self-reliance necessary to secure adequate care here depends a great deal on the level of environmental consciousness in the community in question. To the extent environmental consciousness is lacking, then economic consciousness has to do all the work, so there are many places (such as resource-dependent local economies in the American West) where only autarchy would do the trick.

There is no need in this scheme of things to privilege the nation-state, and every reason not to; few, if any, ecological problems coincide with national boundaries. The institutional possibilities associated with bioregionalism merit further exploration (see McGinnis, 1998). Bioregionalism begins with a rejection of ecologically arbitrary political units, such as counties, provinces, states, and nation-states. Bioregional boundaries are defined instead by watershed, topography, or species composition of ecosystems. But bioregionalism is not just a matter of redrawing political boundaries: it is also a matter of living in place. Redesigned political units should promote, and in turn be promoted by, awareness on the part of their human inhabitants of the biological surroundings that sustain them.

While one can argue endlessly about exactly how and where bioregional boundaries should be drawn, it is easy to see that (say) a bioregional authority for the Colorado River basin makes more ecopolitical sense than dividing control of the basin between six states and a federal government. Among existing bioregional authorities, one of the most interesting is the Northwest Power Planning Council (NWPPC) celebrated by Kai Lee (1993), which is actively engaged in ecological restoration in the Columbia River Basin in the Pacific Northwest of the United States. The NWPPC relies for its ecological information not just upon the work of biological scientists as transmitted to ecological managers, but also on structured opportunities for participation in a variety of forums organized by the Council. These forums welcome the various human users of the ecosystem: Native American tribes, commercial and recreational fishers, other recreational users, wilderness advocates, and those who rely on the river for power, water, and navigation. All these people have material interests of their own to pursue; but many of them are also in day-to-day contact with particular aspects of the ecosystem, and therefore in a much better position than distant managers or politicians to hear news from it. The very existence of these forums in the overtly bioregional context provided by the NWPPC has helped to cultivate bioregional awareness among participants. This is not a matter of discarding material interests in favour of a different set of ecological interests, but rather recognizing how these material interests are aspects of ecological well-being. Such effects are consistent with the capacity of discursive democracy to induce individuals to reflect upon their preferences and so broaden their conception of interests.

Co-ordination Through Spontaneous Order

An ecological democracy could, then, contain numerous loci of political authority, including bioregions. Obviously not all ecological problems and feedback signals reside at the local level. Some of them are global, and hence demand global institutional response. The obvious question here is: how does one co-ordinate between the various sites of political authority, given that one cannot (for example) resolve air pollution problems while completely ignoring the issue of water pollution, or deal with local sulphur dioxide pollution while ignoring the long-distance diffusion of sulphur dioxide in acid rain? The way this co-ordination is currently accomplished is by privileging one level of political organization. In unitary political systems, this will normally be the national state, though matters can be a bit more complicated in federal systems. The state (national or sub-national) will of course often contain an anti-pollution agency that (nominally, if rarely in practice) coordinates policy in regard to different kinds of pollutants. But, as I have already pointed out, from an ecological point of view this is an entirely arbitrary solution, and no more defensible than privileging the local community, or for that matter the global community.

The state and its environmental problem-solving capacities are likely to be with us for the foreseeable future, so their role in coordination should not be ignored. However, an ecological perspective points to kinds of co-ordination that are not organized centrally (as in the state), but arise as emergent properties as the scale of ecological and social organization rises. Such spontaneous orders can achieve coordination where the state does not or cannot—for example, at the system level in international affairs. . . .

The best-known such order is the market, explicitly celebrated in these terms by F. A. von Hayek (see Goodin 1992, 154). But markets are not exactly an ecological success story, as I have already noted. Nor are they much good at co-ordinating the activities of political authorities. Within decentralized political systems, co-ordination is achieved largely through the spontaneous order of partisan mutual adjustment, which to Lindblom (1965) is at the core of collective decision in pluralistic liberal democracies. Such regimes may contain more formal and consciously-designed constitutions, but partisan mutual adjustment proceeds regardless of the content of such formalisms. This adjustment involves a complex mix of talk, strategy, commitment, and individual action devised in response to the context created by the actions of others. As I argued earlier, this kind of spontaneous order under liberal pluralism leaves much to be desired when scrutinized in an ecological light. . . .

Let me suggest that there are two related kinds of spontaneous order which might perform the requisite co-ordinating functions quite well. The first is that which exists in connection with the organizations of civil society. The environmental movement and its associated networks are now international, and organizations such as Greenpeace or Friends of the Earth International can bring home to particular governments the international dimension of issues, such as (say) the consequences to Third World countries of toxic wastes exported by industrialized countries. . . .

To take another example, international public spheres constituted by indigenous peoples and their advocates can bring home to boycotters of furs in London or Paris the resulting economic devastation such boycotts imply for indigenous communities in the Arctic, which rely for cash income on trapping. A public sphere on a fairly grand scale was constituted by the unofficial Global Forum that proceeded in parallel with the United Nations Conference on Environment and Development in Rio in 1992. The point is that the reach of public spheres is entirely variable and not limited by formal boundaries on jurisdictions, or obsolete notions of national sovereignty. And they can come into existence, grow, and die along with the importance of particular issues. So, for example, it is entirely appropriate that the West European peace movement declined as cold war tensions eased in the 1980s.

A second kind of spontaneous order exists in association with the organizations of civil society, but at the same time transcends them. This second kind of order is discursive. Especially when institutional hardware is weak or absent, discourses as social phenomena can and do co-ordinate the understandings and actions of disparate actors. These discourses need know no geographical boundaries. Of course, it matters a great deal how and by whom their terms are set. A dispersed capacity to determine these terms of discourse is especially conducive to coordination across space and across issue areas. . . . Such dispersed capacity finds expression in the network form of organization, itself at home in civil society and the public sphere. The exemplary network is the environmental justice movement, within the United States and across national boundaries. Networks of this sort seem to be particularly at home in the environmental area.

Such networks and related kinds of spontaneous order should not be construed as operating smoothly, consensually, and in fully rational terms. In practice they feature information asymmetries, conflicts, and misunderstandings. In conflicts with other centres of political power, sometimes they will prevail, sometimes not. But in all of these features they are no different from imposed orders in human affairs, such as state bureaucracies, legal systems, or liberal constitutions. Yet out of the negotiation of difference and conflict order can emerge—and disappear. There is no reason to lament the disappearance of a particular network or discourse. Defensible spontaneous orders are problem-driven and do not outlive their usefulness—unlike, for example, state bureaucracies, which are often near-immortal.

There are many different ways of achieving co-ordination in collective decision-making, some spontaneous, some imposed. Examples are hierarchies, markets, bargaining, law, coercion, violence, discussion, partisan mutual adjustment, and moral persuasion. . . . I have emphasized here sources of order that are defensible in both ecological and democratic terms. My argument is not that dispersed control over the contestation of discourses and associated networks should completely replace the more familiar sources, but that the world will be a greener and more democratic place to the extent their relative weight increases.

Conclusion

In contemplating the kinds of communication that might ensure more harmonious co-ordination across political and ecological systems, there is an ever-present danger of lapsing into ungrounded idealism and wishful thinking. I have tried to develop an alternative to ungrounded idealism by showing how discursive democracy can be extended in a direction that overcomes anthropocentric arrogance and that can cope more effectively with the ecological challenge. Democracy is, if nothing else, both an open-ended project and an essentially contested concept; indeed, if debates about the meaning of democracy did not occur in a society, we would hesitate to describe that society as truly democratic. . . .

References

Barry, John. 1999. *Rethinking Green Politics. Nature, Virtue and Progress.* London: Sage.

Bickford, Susan. 1996. *The Dissonance of Democracy: Listening, Conflict, and Citizenship.* Ithaca, NY: Cornell University Press.

Christoff, Peter. 1996. "Ecological Modernisation, Ecological Modernities." *Environmental Politics* 5: 476–500.

Dahl, Robert A. 1956. *A Preface to Democratic Theory.* Chicago: University of Chicago Press.

Dobson, Andrew. 1996. "Democratizing green theory: Preconditions and principles." In *Democracy and Green Political Thought: Sustainability, Rights and Citizenship*, ed. Brian Doherty and Marius de Geus, 132–148. London: Routledge.

Dryzek, John S. 1987. *Rational Ecology: Environment and Political Economy.* New York: Basil Blackwell.

Dryzek, John S. 1990 "Green reason: Communicative ethics of the biosphere." *Environmental Ethics* 12: 195–210.

Eckersley, Robyn. 1992. *Environmentalism and Political Theory: Toward an Ecocentric Approach*. Albany, NY: State University of New York Press.

Eckersley, Robyn. 2000. "Deliberative democracy, ecological risk, and communities-of-fate." In *Democratic Innovation: Deliberation, Association, and Representation*, ed. Michael Saward. London: Routledge.

Ehrenfeld, David. 1978. *The Arrogance of Humanism*. New York: Oxford University Press.

Goodin, Robert E. 1992 *Green Political Theory*. Cambridge: Polity.

Goodin, Robert E. 1996. "Enfranchising the Earth, and its alternatives." *Political Studies* 44: 835–849.

Gore, Albert. 1992. *Earth in the Balance: Ecology and the Human Spirit*. Boston: Houghton Mifflin.

Hajer, Maarten A. 1995. *The Politics of Environmental Discourse: Ecological Modernization and the Policy Process*. Oxford: Oxford University Press.

Janicke, Martin, Helmut Weidner, and Helga Jorgens (eds.). 1997. *National Environmental Policies: A Comparative Study of Capacity-Building*. Springer-Verlag.

Lee, Kai N. 1993. *Compass and Gyroscope: Integrating Science and Politics for the Environment*. Washington, DC: Island Press.

Lindblom, Charles E. 1965. *The Intelligence of Democracy: Decision Making through Mutual Adjustment*. New York: Free Press.

McGinnis, Michael (ed.). 1998. *Bioregionalism*. London: Routledge.

Patten, Bernard C., and Eugene P. Odum. 1981. "The cybernetic nature of ecosystems." *American Naturalist* 118: 886–895.

Plumwood, Val. 1995. "Has democracy failed ecology?" *Environmental Politics* 4: 134–168.

Weale, Albert. 1992. *The New Politics of Pollution*. Manchester: Manchester University Press.

20

Toward the Regional*

Gar Alperovitz

Like Schumacher, Thayer, Dryzek, and others in this book, political scientist Gar Alperovitz raises the critical question of scale. *For Alperovitz the issue is the feasible scale of a democratic nation, particularly the United States. To understand the importance of getting the scale right, Alperovitz discusses the negative effects of large-scale, centralized power on democracy. This book adds other concerns—fossil fuel dependence, defensive expenditures for environmental disruption, pay-down of accumulated financial, infrastructural, and natural debts—that constitute new challenges to large-scale democratic governance. When considered as a whole, these concerns evoke a powerful argument for decentralization. Also notice the evidence Alperovitz presents suggesting that significant decentralizing trends are already underway in the United States and elsewhere.*

Is it really feasible—in systemic and foundational terms—to sustain [equality, liberty, and democracy] in a very large-scale, centrally governed continental system that spans almost three thousand miles and includes almost 300 million people? And if not, how might a democratic nation ultimately be conceived?

Reflection on the impact of very large scale on democracy can be traced back to the Greeks, and later especially to Montesquieu, who held that democracy could flourish only in small nations. The judgment that very large scale is inimical to democracy was also taken very seriously by the founding fathers. Indeed, at a time when the United States hardly

*Alperovitz, Gar. (2004). "Democracy: Is a Continent Too Large?" and "The Regional Restructuring of the American Continent." In *America beyond Capitalism: Reclaiming Our Wealth, Our Liberty, and Our Democracy*, 63–69 and 152–166. Hoboken, N.J.: John Wiley & Sons.

extended beyond the Appalachian mountains, John Adams worried: "What would Aristotle and Plato have said, if anyone had talked to them, of a federative republic of thirteen states, inhabiting a country of five hundred leagues in extent?" Similarly—again, at a time when the nation numbered a mere 4 million people—even James Madison (who challenged the traditional argument that democracy was possible only in small nations) believed that a very large (rather than a "mean"-scale) republic could easily become a de facto tyranny because elites at the center would be able to divide and conquer diverse groups dispersed throughout the system. Few people imagined democracy in a continent.

One can also isolate important and difficult aspects of the question of scale in the larger complex of issues that in the nineteenth century culminated in the Civil War. For our purposes, however, it is sufficient to recall that a sophisticated theoretical debate over scale problems began to develop in academic and political centers during the early years of the twentieth century, continuing up to and through the 1920s and 1930s.

The traditional response to the argument that democracy is difficult if not impossible in very large scale units has been to propose decentralizing to the states. The point of departure for the more sophisticated debate is recognition that many states are simply too small to manage important economic issues, or for instance (in the 1930s as well as in modern times) a number of important ecological matters. Logically, if a continental national system is too large and many states are too small, the obvious answer must be somewhere in-between—the unit of scale we call a "region."

Historian Frederick Jackson Turner (1932, 45; 194) put it this way: "There is a sense in which sectionalism is inevitable and desirable"—going on to observe: "As soon as we cease to be dominated by the political map, divided into rectangular states . . . groups of states and geographic provinces, rather than individual states, press upon the historian's attention."

A leading conservative theorist who urged the same logic during the 1930s was Harvard political scientist William Yandell Elliot (1935, 193): "Regional commonwealths would be capable of furnishing units of real government, adequate laboratories of social experiment, and areas suited to economic, not-too-cumbersome administration."

"The libertarian argument against 'too much government,'" Henry C. Simons (1948, 12; 21) held, "relates mainly to national governments, not to provincial or local units—and to great powers rather than to small nations." Simons believed a "break-up" of the United States "desirable" (though he did not think it politically feasible). The alternative, he urged, required taking seriously a process of "steady decentralization."

Another Harvard professor and president of the American Political Science Association, William Bennett Munro (1928, 137; 153–154), did not mince words about the issues and logic that he believed needed to be confronted: "Most Americans do not realize what an imperial area they possess," he said, adding, "Many important issues and problems . . . are problems too big for any single state, yet not big enough for the nation as a whole. . . . They belong by right to regional governments."

The then innovative experiment with regionalism, environmental management, and public ownership—the Tennessee Valley Authority—was related to the early twentieth-century regional rethinking movement, as were proposals by Franklin Delano Roosevelt for similar regional authorities throughout the country. Even though President Harry Truman continued to offer such ideas (in connection, for instance, with the Columbia River), the regionist movement was cut short by a combination of anti–New Deal politics and the advent of World War II and the era of the Cold War.

The underlying issue of scale, however, is plainly still with us. Moreover, the extraordinary cost of modern campaigning in large areas has added to the advantages that scale gives to wealthy elites and corporations, thereby further undermining democratic possibilities. Money talks louder when expensive television ads are the only way to reach large numbers. During the election of 2000 an estimated $1 billion was spent on television alone. . . .

We rarely pause to consider the truly huge size of the American system. The fact is, the United States is extreme in scale. Germany could fit within the borders of Montana alone; France is smaller than Texas. . . .

A 1973 book, *Size and Democracy*, by Robert A. Dahl and Edward R. Tufte . . . concluded [that] "the inexorable thrust of population growth makes a small country large and a large country gigantic" (p. 2).

Daniel Bell (1987, 14) [argued] that the "nation-state is becoming too small for the big problems of life, and too big for the small problems of life." Bell went on, "[T]he flow of power to a national political center means that the national center becomes increasingly unresponsive to the variety and diversity of local needs. . . . In short, there is a mismatch of scale."

George F. Kennan (1993, 143; 149) took the argument a step further: "We are, if territory and population be looked at together, one of the great countries of the world—a monster country. . . . And there is a real question as to whether bigness in a body politic is not an evil in itself, quite aside from the policies pursued in its name." Kennan proposed long-term regional devolution which, "while retaining certain of the rudiments of a federal government," might yield a "dozen constituent republics, absorbing not only the powers of the existing states but a considerable part of those of the present federal establishment."

The radical historian, the late William Appleman Williams (2000), suggested embodying socialist principles: "[T]he issue is not whether to decentralize the economy and the politics of the country, but rather how to do so. The solution here revolves about the regional elements that make up the existing whole." And the modern conservative regionalist, Donald Livingston, asked in 2002: "What value is there in continuing to prop up a union of this monstrous size?" He went on: "[T]here are ample resources in the American federal tradition to justify states' and local communities' recalling, out of their own sovereignty, powers they have allowed the central government to usurp." . . .

A converging trend of environmental thinking is also significant. Much of the work of the Environmental Protection Agency is already organized along regional lines, and as Harvard analyst Mary Graham observes, "Many environmental problems are inherently regional in scope, rather than national or local." Among the interesting proposals here are suggestions by Kirkpatrick Sale that emphasize small "bio-regions" and work on "eco-regions" by World Wildlife Fund experts in the Conservation Science Program.[1] A fully developed long-term ecological vision in which many "regions within the United States could become relatively self-sufficient" has been put forward by Herman Daly and John Cobb: "[T]he nation-state is already too large and too remote from ordinary people for effective participation to be possible" (1989, 179; 293).

The various developing arguments have also received indirect support from research on the achievements of smaller-scale nations—some roughly equivalent to U.S. regions. The Scandinavian nations and such countries as Austria and the Netherlands, for instance, have demonstrated that smaller scale has commonly helped—rather than hindered—their capacity to deal with major economic, social, and environmental problems. In general, equality has been greater and unemployment rates lower than in most larger European countries. Moreover, although such nations are far more involved in trade than larger nations, for the most part they have found more effective ways to manage the dislocations and other challenges brought about by economic globalization.

In a related development, political scientist Michael Wallerstein has demonstrated how the United States' very large scale and huge labor force have made union organizing both difficult and expensive. This in turn has tended to limit union size, thereby both weakening collective bargaining and undercutting one of the primary organizational foundations of progressive political-economic strategies in general.

The modern reemergence of regionalist ideas is no accident. Although the primary thrust of the argument concerns what it takes to achieve democratic accountability and participation, the American discussion is part and parcel of a world-wide trend that is already producing different forms of regional devolution in nations as diverse as Britain and Canada, on the one hand, and China and the former Soviet Union, on the other. . . .

A global perspective, in fact, suggests that the quietly emerging—and seemingly unusual—American arguments are only the beginning of something that is all but certain to grow in force and sophistication as time goes on. . . .

The pluralist commonwealth model attempts to deal seriously with long-standing arguments that the sheer continental size of the United States and its very large population are ultimately inimical to a robust system-wide vision of democratic practice. . . . Community-oriented strategies appear to be within the range of realistic political possibility in coming years. What of the larger and seemingly utopian idea that much more far-reaching—indeed, radical—decentralization is both necessary and possible?

Five major considerations suggest that, contrary to conventional assumption, the logic of regional restructuring is likely to become of increasing importance as the twenty-first century develops. These include trends in Supreme Court and congressional decision making; an explosion of state based initiatives; the impact of global political-economic forces on the current federal system; very large-order projected changes in the economy and population; and new trajectories of expanding ethnic political power concentrated in key regions experiencing economic distress.

Over the last several decades a series of Supreme Court and congressional decisions has begun to establish new principles of decentralization in the U.S. federal system that (for better or worse) are much more far-reaching than many understand. At the same time, numerous states have launched new initiatives that are slowly altering the locus of power in the system.

The trend in Supreme Court decision making has been well documented. . . . In *Seminole Tribe of Florida v. Florida* and several subsequent cases involving state employees, savings banks, and violence against women, the Court held that Congress did not have authority to establish federal jurisdiction over states that did not consent to be sued. . . .

An equally important trend in federal legislative actions has furthered the decentralization process. . . . Moves in the direction of greater state and local authority . . . [for example] . . . allow great latitude in the use of federal money for various urban housing and community development programs. . . .

Similarly, the Intermodal Surface Transportation Efficiency Act gives states considerable discretion in developing transportation programs in accord with local priorities. . . . Innovative and widely publicized health insurance strategies in Oregon, Vermont, Hawaii, and Maine, among others, have [also] been developed on this basis.[2]

The movement toward greater state authority is not an unbroken trend. A countermovement is evident in several Supreme Court decisions related to economic issues and in legislative efforts to enact "preemption clauses" mandating federal jurisdiction in connection with various regulatory matters. On the other hand, the state attorneys general have mounted important new legal challenges, most dramatically in connection with tobacco, but also with regard to inflated costs of prescription drugs, antitrust (Microsoft), and other issues ranging from securities fraud to global warming. . . .

Independent legal activism by the states has also arisen in large part because of federal inaction. Modern state attorney general initiatives first began to develop in response to the Reagan Justice Department's failure to do much to protect consumers and the environment. The $206 billion tobacco settlement in 1998 was a major victory that helped put the general movement into high gear.

In general, University of Virginia political scientist Martha Derthick points out, the states have increasingly become the "default setting" of the American political-economic system—the level of government that acts when Washington does not because of gridlock or neglect. Alan Ehrenhalt (2002, 6–8) of *Governing* magazine goes further: states are now increasingly the "level of government we go to because we don't expect the others to succeed."

Many traditional liberals, fearing a weakening of federal standards, have opposed the general trend. Others feel the only option available may be a long-haul effort to rebuild power at the base, state by state.

The important point for the future, Ehrenhalt (2002, 6–8) emphasizes, is that "once states and their elected leaders begin thinking of themselves as the actors of first resort on crucial questions—rather than the actors of last resort—the logic of the whole system is in for a change."

The implications of globalization reinforce this fundamental judgment. Especially significant are pressures that create new Washington-level restrictions on state decision making—and in turn produce new and angry resistance. A recent study by Columbia University professor Mark Gordon (2001, 34) of the implications of World Trade Organization (WTO) regulations points out that WTO rules "strike at the heart of the types of policy decisions that States use to define some of their most basic beliefs." WTO regulations now increasingly challenge traditional state prerogatives in connection with "issues of environmental and consumer protection, set asides to assist minority or small businesses, efforts to regulate the activities of large financial services institutions such as banks and insurance companies, and decisions about how to structure the raising of revenue through taxes and its expenditure through government procurement policies."

Gordon (2001, 34) and other analysts predict that as the impact of the new global trade regime hits home, an intense dynamic will be set loose that will force Washington to reach ever deeper into state power to enforce global agreements—and will, in turn, force states to develop ever more adamant counterstrategies: "[G]lobalization introduces a whole series of 'shocks' to the existing system."

Numerous state leaders throughout the country have, in fact, already gone on record challenging WTO and NAFTA imposed requirements. A resolution passed by the Oklahoma legislature—to cite only one of many examples—demands that the president and Congress "preserve the traditional powers of state and local governance" and "ensure that international investment rules do not give greater rights to foreign investors than United States investors enjoy under the United States Constitution."[3]

The long-term logic points to an ever more powerful "backlash" by the states—and demands for greater independence from the long arm of Washington in its role as enforcer of WTO rules. . . .

The likelihood of structural change in the federal system over the course of the century is intimately related to even more fundamental shifts—above all, to emerging economic and population trends. . . .

Twenty-one states have populations of less than 3 million (of these, seven have less than a million). Another nine have populations of less than 5 million. Most of these thirty states (and perhaps others) are too small to deal effectively with many economic, environmental, transportation, and other problems on their own.[4]

Long-term federal restructuring that might ultimately come to rest on a unit of scale larger than most states but smaller than the nation—the region—most likely would begin with states that: (1) are themselves very large; (2) have a sense of their own political and policy identity; (3) are experiencing trajectories of growing racial and ethnic change different from the rest of the nation; (4) are experiencing particularly painful economic and fiscal distress; and (5) are already constituted as organized "polities."

An obvious candidate to initiate long-range change is the regional-scale "mega-state" of California. California, in fact, is already the equivalent of a very large semiautonomous political-economic system. Its economy is roughly the size of France's, the fifth-largest economy in the OECD. The economy of the five-county Los Angeles area alone is roughly the size of Spain's, the OECD's ninth largest economy—and is greater than the economies of Brazil, India, and South Korea (Los Angeles County Economic Development Corporation 2002).

California's population of 35 million is greater than that of Canada (31 million), Australia (19 million), the Netherlands (16 million), Portugal (10 million), and all four of the Scandinavian countries combined (24 million). Los Angeles County is larger in population than forty-two of the fifty states. The state is also larger, geographically, than numerous important nations—including Germany, Japan, the United Kingdom, Poland, and Italy (U.S. Census Bureau n.d.; *OECD in Figures: Statistics on the Member Countries* 2002, 6–7; U.S. Census Bureau 2002). . . .

. . . California has also already been impacted by other globalization pressures, and numerous of its state programs are likely to run afoul of WTO and NAFTA regulations. Its massive fiscal problems—and recent electoral events—suggest the likelihood of ongoing political volatility. Given its economic difficulties and the emerging pressures, in many ways it would be surprising, in fact, if a large and inherently wealthy regional-size state like California did not at some point *demand* greater powers to better manage its own affairs.

If (when?) it did, its example would likely be followed in one way or another by other large states. Texas, which now numbers 20.9 million, is projected to reach 27.2 million by 2025 and, on reasonable assumptions, 46 million by century's end. . . . Florida and New York are also of substantial interest. Florida is larger geographically than many midsize European countries; its current 15.9 million population is projected to reach 35 million by 2100. New York's population of 18.9 million could reach 33.5 million and its economy grow to over $8 trillion by 2100. All three states might follow the lead of California—or at some point launch independent initiatives of their own that would have repercussions throughout the system.

Other plausible decentralization scenarios involve groups of smaller states. Numerous precedents and a long history of states working together could be drawn upon either in response to an assertion of power by larger states or simply in order to achieve positive goals that few small states can achieve on their own. Regional strategies have long been common, for instance, in connection with environmental issues. Some regions, such as New England, have developed multiple forms of interstate cooperation involving groupings of governors, attorneys general, environmental administrators, and others.[5]

Nearly two hundred Interstate Compacts—which are already authorized by the Constitution—also currently coordinate various state efforts in connection with matters ranging from economic development to high-speed intercity passenger rail service. Federal precedents also abound—including the Tennessee Valley Authority and previously noted presidential proposals for many similar authorities. . . . The Appalachian Regional Commission currently involves some thirteen states in common efforts related to industrial development, energy resource coordination, tourism promotion, and other matters. Both the Johnson and Nixon administrations experimented with various additional forms of regionalization—the former by establishing regional commissions, the latter through regional administrative strategies.[6]

Such precedents for regional coordination do not reach to the many larger issues of political-economic authority and power that system-wide restructuring would clearly require. On the other hand, the historical record offers evidence that states working together when problems are larger than any one state can handle have been effective in many, many instances. The regular reappearance of regional efforts also points to a certain political appeal that regionalist ideas appear to have—especially when traditional alternatives are incapable of dealing with pressing political-economic problems.

Few in the United States are aware that in recent decades an intense exploration of regionalist constitutional changes has been under way throughout the world—in Britain and in nations as diverse as China, Italy, Indonesia, the former Soviet Union, and Canada. In 1989 a comprehensive international report concluded that decentralization had become a "subject of discussion in all countries regardless of whether they are old or young states or whether they have a long unitary or federal tradition"(Konig 1989, 3).

It is possible that the United States will be immune to the global trend—and that as the nation moves toward 500 million and beyond, it will continue to be managed, administered, and fundamentally governed from Washington without significant change in what by century's end will be a constitutional structure that is more than three hundred years old. However—and even though few Americans have yet imagined the possibility—given the various changes under way, the odds are that population growth alone will ultimately create conditions that demand consideration of some form of major restructuring.

The specific shape a new Pluralist Commonwealth oriented regionalism might take over the course of the century is obviously indeterminate. Initial changes would likely involve greater state/regional autonomy in connection with economic and environmental matters, reductions in federal preemptive powers with regard to corporate regulation, limitations on the impact of WTO and other trade treaties on state/regional legislative authority, and alterations in current Constitutional Commerce Clause restrictions related to state/regional economic rights. Beyond this, much larger issues concerning the apportionment of power might well be posed. . . .

. . . Quite apart from population and other pressures that may force change—and the many uncertainties that would ultimately have to be confronted and resolved—over the long arc of the twenty-first century, Americans who are committed to a renewal of democracy are unlikely to be able to avoid the truth that in all probability this can only be meaningfully achieved in units of scale smaller than a continent but also of sufficient size to be capable of substantial semiautonomous functioning: the region.

Notes

1. Graham 1998, 63–60, 70; Sale 1985 and World Wildlife Fund. *Ecoregions project snapshot.* Retrieved 04/09/03 from www.worldwildlife.org/ecoregions/related_projects.htm.

2. Donahue 1997, 8; Dilger 1998, 49; Schneider 1997, 89; Greenhouse 2003.

3. A partial list of resolutions and letters of protest—including the Oklahoma State Legislature's resolution—can be found at Public Citizen, "State and Local Opposition to NAFTA Chapter 11," retrieved 31 October 2002 from http://www.citizen.org/trade/nafta/CH__11/articles.cfm?ID=7619.

4. State populations as of the 2000 Census, retrieved 2 October 2003 from http://quickfacts.census.gov/qfd/.

5. See, for example, Fargin 1999; Allen 1998; Gavin 2001; and Johnson 2001.

6. Council of State Governments, "Compacts Believed to Be in Effect in 2001," retrieved 29 October 2002 from http://ssl.csg.org/compactlaws/comlistlinks.html. For federal regional precedents, see Martha Derthick, *Between State and Nation* (Washington, D.C.: Brookings Institution, 1974); and Ann Markusen, *Regions: The Economics and Politics of Territory* (Totowa, N.J.: Rowman & Littlefield, 1987). For information on the continuing work of the Appalachian Regional Commission, see Appalachian Regional Commission, retrieved 4 November 2002 from http://www.arc.gov/index.jsp.

References

Allen, Scott. 1998. Cut is OK'd in emissions of mercury; New England governors join Canadian premiers in accord. *Boston Globe,* June 9, A1.

Arrandale, Tom. 2002. The pollution puzzle. *Governing* 15:22–26.

Bacon, John and Haya El Nasser. 2000. Vermont governor signs gay-union bill. *USA Today,* April 27, A3.

Balchin, Paul, Luděk Sýkorla, and Gregory Bull. 1999. *Regional policy and planning in Europe*. London: Routledge.

Baldassare, Mark. 2000. *California in the new millennium*. Berkeley: University of California Press and Public Policy Institute of California.

Bell, Daniel. 1987. The world and the United States in 2013. *Dædalus* 116:14.

Claiborne, William. 1995. Wilson challenges Hill to match his hard line. *Washington Post*, January 10, A7.

Cushman, John H. Jr. 1995. Congress limits federal orders costly to states. *New York Times,* February 2, A1.

Daly, Herman E., and John B. Cobb. 1989. *For the common good*. Boston: Beacon Press.

Dilger, Robert Jay. 1998. TEA-21: Transportation policy, pork barrel politics, and American federalism. *Publius: The Journal of Federalism* 28: 49–69.

Donahue, John D. 1997. *Disunited states*. New York: Basic Books.

Dorman, Robert L. 1993. *Revolt and the provinces: The regionalist movement in America, 1920–1945*. Chapel Hill: University of North Carolina Press.

Dubinsky, Paul R. 1994. The essential function of federal courts: The European Union and the United States compared. *American Journal of Comparative Law* 42.

Ehrenhalt, Alan. 1999. Demanding the right size government. *New York Times,* October 4, A27.

Ehrenhalt, Alan. 2002. The Monkey or the Gorilla. *Governing* 15: 6–8.

Elliott, William Yandell. 1935. *The need for constitutional reform: A program form national security.* New York: McGraw-Hill.

Fagin, Dan. 1999. Ill winds blow: As progress on clean air stalls, regional fights and attacks on rules intensify. *Newsday*, October 17, A7.

Freudenheim, Milt and Melody Petersen. 2001. The drug-price express runs into a wall. *New York Times*, December 23, C1.

Gavin, Robert. 2001. States rediscover energy policies. *Wall Street Journal*, March 21, B13.

Goggin, Malcolm L. 1999. The use of administrative discretion in implementing the state children's health insurance program. *Publius: The Journal of Federalism* 29: 35–51.

Gordon, Mark C. 2001. *Democracy's new challenge: Globalization, governance, and the future of American federalism*. New York: Demos.

Graham, Mary. 1998. Why states can do more. *The American Prospect* 9: 63–60; 70.

Greenhouse, Linda. 2002. Court, 5–4, upholds authority of states to protect patients. *New York Times*, June 21, A1.

Greenhouse, Linda. 2003. Justices allow drug-cost plan to go forward. *New York Times*, May 20, A1.

Hakim, Danny. 2002. At the front of pollution. *New York Times*, July 3, A1.

Johnson, Kirk. 2001. A changing climate in ideas about pollution. *New York Times*, May 20, 39.

Keating, Micheal and John Loughlin, eds. 1997. *The political economy of regionalism*. London: Frank Cass & Co.

Kennan, George F. 1993. *Around the cragged hill: A personal and political philosophy*. New York: W.W. Norton.

Konig, Klaus. 1989. Appraisal of National Policies of Decentralization and Regionalization. Report for Research Committee I: Law and Science of Public Administration, International Institute of Administrative Sciences, XXIst International Congress, Marrakech, 3.

Landsberg, Mitchell and Miguel Bustillo. 2001. Davis says all power costs to be recovered. *Los Angeles Times*, April 14, A1, A21.

Los Angeles County Economic Development Corporation. 2002. Press Release, *2002–2003 Southern California Five-County Area Economic Forecast and Industry Outlook* (September 16). Retrieved 29 October 2002 from http://www.laedc.org/data/press/PR65.shtml.

Luccarelli, Mark. 1995. *Lewis Mumford and the ecological region: The politics of planning*. New York: The Guilford Press.

Mooney, Chris. 2001. Localizing globalization. *The American Prospect* 12: 23–26.

Moore, Harry Estill. 1937. *What is regionalism?* Chapel Hill: University of North Carolina Press.

Morain, Dan and Richard Simon. 2001. Energy deal may take a month, Davis tells analysts. *Los Angeles Times*, March 1, A3, A18.

Munro, William Bennett. 1928. *The invisible government.* New York: Macmillan.

Nieves, Evelyn. 2000. California gets set to shift on sentencing drug users. *New York Times* November 10, A18.

Odum, Howard W. and Harry Estill Moore. 1996. *American regionalism: A cultural-historical approach to national integration.* Gloucester: Peter Smith.

OECD in Figures: Statistics on the Member Countries. 2002. Paris: OECD.

Perlman, Ellen. 1999. Rail's resurgence. *Governing* 12: 28–30.

Public Citizen. 2001. *NAFTA Chapter 11 investor-to-state cases: Bankrupting democracy—Lessons for fast track and the free trade area of the Americas.* Washington, D.C.: Public Citizen. (September).

Sale, Kirkpatrick. 1985. *Dwellers in the land: The bioregional vision.* San Francisco: Sierra Club Books.

Schevitz, Tanya, Lori Olszewski, and John Wildermuth. 2000. New demographics changing everything. *San Francisco Chronicle*, August 31, A1.

Schneider, Saundra K. 1997. Medicaid section 1115 waivers: Shifting health care reform to the states. *Publius: The Journal of Federalism* 27: 89–109.

Simons, Henry C. 1948. *Economic policy for a free society.* Chicago: University of Chicago Press.

Steiner, Michael and Clarence Mondale. 1988. *Region and regionalism in the United States: A source book for the humanities and social sciences.* New York: Garland Publishing.

Strossen, Nadine. 1990. Recent U.S. and international judicial protection of individual rights: A comparative legal process analysis proposed synthesis. *Hastings Law Journal* 41.

Sykes, Leonard Jr. 2001. Attacking the cause of crime: Group pushing drug treatment instead of prison for offenders. *Milwaukee Journal-Sentinel,* December 28, B1.

Tanenwald, Robert. 1998. Implications of the balanced budget act of 1997 for the "Devolution Revolution." *Publius: The Journal of Federalism* 28: 23–48.

Turner, Frederick Jackson. 1932. *The significance of sections in American history.* New York: Henry Holt.

U.S. Census Bureau. 2002. *Radio zone quotes and sound* bites: *County population estimates and rankings.* Retrieved 23 April 2003 from www.census.gov/pubinfo/www/radio/sb_2002countypopest.html.

U.S. Census Bureau. *State and county quick facts*: *California.* Retrieved 31 October 2002 from http://quickfacts.census.gov/qfd/states/06000.html.

Walters, Jonathan. 2001. Save us from the states! *Governing* 14: 20–27.

Wei, Yehua. 1998. Economic reforms and regional development in coastal China. *Journal of Contemporary Asia* 28: 498–517.

Williams, William Appleman. 2000. *Empire as a way of life.* Oxford: Oxford University Press.

21

Global Problems, Localist Solutions*

David J. Hess

Science and technology scholar David J. Hess sees localist movements arising in the United States and around the world in response to the failings of globalization and to the power of transnational corporations. He argues that the movements must be understood on their own terms—namely, small businesses and nonprofits, locally and independently owned, that have a vision of building an alternative global economy (see the box titled "Energy Islands"). For these movements, justice resides in the improved economic and political autonomy of place-based communities.

The driver of localist movements based in the small-business sector is the shortcomings of an economy dominated by huge companies. Local ownership opens new economic and political space. Combined with the premises of this book, the decline in net energy and waste sinks, and, say, Sale's history of decentralization, Thayer's notion of bioregion, and Litfin's ecovillages, we get a vision of social change constrained by biophysical reality but, if we look carefully, also of expanding social space.

Mainstream political debates are based on the hope that the global economy can simultaneously undergo greening and continued economic growth without destroying the environment or plunging the world's poor into epidemics and starvation. The hope is based on the assumption that technological innovation can be rapid enough to compensate for the environmental impact that accompanies increased economic growth, and that governments can provide adequate policy solutions before the

*Hess, David J. 2009. "Global problems and localist solutions." In *Localist Movements in a Global Economy: Sustainability, Justice, and Urban Development*, Cambridge, MA: MIT Press, 23–67. Modified by the author and reprinted with permission.

Energy Islands

In just one decade, Samsø—a small Danish island community of 43,000—transformed the base of its energy consumption from oil and coal to renewable sources. Located in the North Sea, Samsø once imported all of its energy from the mainland via tankers and electric cables. Now, enough renewable energy is generated on the island itself not only to power the entire island, but also to export 10 percent to the mainland.

The island's location does produce an offshore wind advantage, but other renewable options such as straw burning, solar energy, and biofuels are also utilized. Wind turbines, some purchased individually, others collectively, generate yearly dividends from the energy produced. Citizens have also built small-scale wind turbines in their backyards, adopted straw-burning furnaces, and begun heating domestic water with solar collectors.

What makes this story surprising is how ordinary the citizens are. There were no early adopters of renewable energy on the island. Rather, the transformation started only after an engineer who did not live on the island entered a contest held by the Danish Ministry of Environment and Energy aimed at generating plans for reducing Denmark's fossil fuel dependence. Based on his winning plan, the resulting project supported just one staff person. Despite this, the project cultivated widespread support by stressing the role of the whole community. When the renewable energy implementation project finally became a reality, it had wide participation. Now, support for renewable energy is the norm and Samsø is of great interest to those who study durable living.

Source: Elizabeth Kolbert, "The Island in the Wind," *New Yorker*, July 7 and 14, 2008, 68–77.

catastrophic environmental effects of ongoing economic growth and ecological collapse are widely felt.

In defense of that hopeful view, it should be noted that since the 1970s the world's national governments have constructed environmental agencies and programs, and they have also increasingly recognized the importance of responding to various global environmental crises, such as climate change, fishery depletion, water shortage, persistent chemical pollutants, and habitat destruction. Notwithstanding the gains, the mainstream scenarios of a smooth transition to economic growth or steady state within ecological limits face severe shortcomings. One limitation involves the sincerity and pace of the greening of industry. On a first impression, the ostensible greening of large corporations appears to be a hopeful sign of a transition toward the scenario of dematerialization and sustainable production. However, when one looks a little more carefully at the actions of even the greenest of corporations, the record is often more complicated. Even the business press has recognized the difficulties that corporate environmental officers face when attempting to gain support for green innovations that do not have an equivalent return on investment to other investment options. Likewise, more detailed studies of social scientists show an ongoing conflict between profitability goals and public pronouncements of social and environmental responsibility.[1]

A deeper weakness in mainstream nostrums is the failure of corporate greening to lead to a decline in absolute environmental impact. To date corporate greening has coincided with continued growth in absolute levels of resource consumption and environmental degradation at a global level. For example, automobile companies have continued to develop fabulous green-concept vehicles and a new generation of hybrid, electric, and flex-fuel vehicles, but they also compete to put increasing numbers of cars and trucks on the roads rather than envisioning a transition to intensive use of public transportation. Likewise, the big-box retailers are greening their stores and their product lines along the best eco-efficiency principles, but they continue to construct global commodity chains that require increasing amounts of fossil fuel energy for transportation. The electrical utilities are building some wind farms and offering some energy conservation measures, but their revenues remain tied to increased electricity consumption, much of which, in the United States, is based on coal and natural gas. Some oil companies are diversifying to reposition themselves as energy companies, but their profits remain linked to increased petroleum consumption, and they continue to compete with each other to explore and exploit new oil fields all over the world. In short, when attention is placed on the greening of industry, it is too easy to lose track of the fundamental driver of the environmental crises: a global economy that is continuing to grow in absolute levels of environmental sinks and withdrawals. Instead, the focus of attention shifts to incremental innovations that reduce impacts but ignore the difficult politics of the environmental impact of continued growth.

Although there is a greening process underway, it tends to occur too slowly to address the world's looming environmental crises. Among the primary factors behind a deep and lasting ecological transition of the economy is the political influence of brown corporations that benefit most from environmental degradation. Even within so-called green corporations, there are groups whose profitability growth is threatened by new regulatory proposals. As a result, the policymaking process will continue to involve conflicts between relatively green and brown segments of industry and society. To the extent that solutions emerge from the political process, they are likely to be piecemeal and watered down. Because the solutions are likely to be diluted in comparison with what needs to be done to bring about the high degree of dematerialization of the economy that would allow economic growth to occur within environmental limits, critics of mainstream political scenarios envision a much less rosy future: an ongoing environmental crisis and an uneven, decades-long, historical transition to societal collapse.[2]

Collapse will mean many things to many people. In a world of increasing natural disasters, food and water shortages, and other forms of climate-related risk, the wealthy have much less to lose than do the poor, and indeed they have much to gain. Elites have the financial resources to diversify their wealth, insure their investments against risk, and get out of harm's way when the disasters strike. The more conservative segments of the elites, those who support the neoliberal dream of dismantling the public sector, have also begun to find new economic opportunities in a world of privatized disaster relief. A halting policy process of taking one step backward on solving environmental problems, followed by one step forward, provides the wealthy with all sorts of economic opportunities to benefit from both the greening of the economy and the unraveling of the ecology, at least in the short term, which is the only time horizon for the publicly traded corporations in which they are invested. It is likely that the mainstream political field, with its mixes of aggressive neoliberalism and timid social liberalism, will provide an ongoing mixture of half-hearted responses that lead to uneven collapse, environmental degradation, and human immiseration throughout the world.[3]

Localist Alternatives

There are many ways to address the problem behind the problem—that is, the failure of political and economic institutions to respond adequately and rapidly to the environmental crises. Probably the most likely driver of a fundamental technological transition will be the slow growth of the clean-tech sectors of industry, which at some point will provide an effective countervailing power to the "brown" or antienvironmental segments of industry and the state. Another driver of change is the ongoing pressure from social movements that mobilize popular support for regulatory and legislative reform and, on the far left, even public ownership of some industries. Yet a third and less well understood possible driver of change is "localism," understood here as the movement of movements in support of government policies and economic practices oriented toward enhancing local democracy and local independent ownership of the economy in a historical context of corporate-led globalization. Examples of localist movements include the "buy local" and "local living economies" movement of locally owned independent businesses; farmers' markets, community gardens, small farms, and other elements of local food networks; community finance and community media; and various efforts to develop locally owned energy and transportation systems.

The localist movement of the early twenty-first century in the United States can be traced back to various historical sources, including the populist and progressive political movements in the United States as well as the broader political history of anarchist and socialist politics. A more proximate source is the author of *Small Is Beautiful*, E. F. Schumacher, who became a popular figure during the 1970s. Schumacher was a socialist who spent most of his career working as an economist for the National Coal Board, an organization that controlled the United Kingdom's nationalized coal industry. From that vantage point he was able to see the limitations of government ownership of industry and the unflattering similarities between large publicly owned companies and large publicly traded corporations. His thinking also drew on his experiences as a director on the board of an employee-owned company, an economic advisor for the country of Burma, an organic gardener, and a student of the Gandhian, village-centered strategy of rural development. Those experiences came together in his advocacy of a transition to economies based on renewable resources, people-centered and employee-owned business organizations, and technologies of development appropriate to the needs of a country's poor and working-class people. He concluded that neither the large, publicly traded corporation nor the large, government-owned corporation were necessarily the best solutions for building a more socially just and environmentally sustainable society. Instead, he sought answers in new forms of economic organization and ownership.[4]

Although the thought of E. F. Schumacher and others like him who influenced the post-1960s counterculture may be a good starting point for understanding the recent history of localism, there are significant differences between his vision of an alternative economy and that of present-day localists. For today's localist movement in the United States, the emphasis on appropriate technology has been replaced by a more general concern with sustainability and community, and likewise the organizational focus is much more on small businesses than on employee-owned firms. If one wishes to push the comparison, it may be best to think of present-day localism as Small Is Beautiful 2.0, this time with an economic base in a preexisting economic class and with greater concern for independent ownership than appropriate technology. Even that qualified comparison should not be pushed too far, because the class basis of present-day localism is considerably different from that of "small is beautiful" economics, which remained rooted in a vision of building appropriate organizations and technologies for the world's working class and poor people.

Although one can identify prolocalist individuals who are influenced by Schumacher and other political thinkers in the socialist or communalist tradition, it would be a mistake to position localism merely as a continuation of radical political thought associated with socialism and anarchism. Instead, one can identify affinities between localism and various political ideologies. To the radical side, the support of locally owned public enterprises and employee-owned enterprises resonates with socialism. But there are also wings of the localist movement that draw on the radical heritage of decentralization and communalism. In this sense one might classify localism as a continuation of radical political traditions and debates. However, strands of mainstream political thought also are evident in the localist movement. For example, localism is consistent with the neoliberal trend in favor of the devolution of national government responsibilities to the states and to communities. A focus on local governance has flourished in the neoliberal climate of government-driven devolution and privatization. Furthermore, by asking consumers to support locally owned independent businesses, "buy local" campaigns, and other localist mobilizations, advocates of localism work through the market under the consumerist logic of voting by spending. But against this neoliberal strand one can also find strands of thought and policy advocacy that would be better characterized as social liberalism. For example, localist campaigns can also involve local government regulatory interventions and calls for policy support from the federal government, both of which are consistent with the tradition of twentieth-century social liberalism.[5]

The continuities of localism with socialist, communalist, neoliberal, and social liberal politics should all be recognized, and likewise any attempts to reduce localism to one or the other strands of political ideologies would best be greeted with questions about oversimplification. It is too easy for analysts who have sympathies with positions within the existing political field of mainstream and radical politics to misinterpret localism as small-scale socialism or social liberalism, an iteration of the communalist politics of the late 1960s and early 1970s, or an expression of neoliberalism via marketplace reformism. Rather, if one starts with recognition of the diversity of the localist movement, it becomes possible to recognize the types of coalitions that are being built at the grassroots level and to explore both the continuities with and differences from political legacies. Localism can appeal to socialists who want to see more local government ownership; to communalists and decentralists who wish to see the growth of independent local economies; to neoliberals who support the small-business sector as a solution to social and environmental problems; and to social liberals who seek greater regulation of local land use and federal legislation that ends corporate handouts. The bluest of Democrats may find themselves agreeing with the reddest of Republicans, at least on the strategy of local economic control as a means for improving the environmental, health, and quality of life of their shared, place-based communities. Furthermore, the selection of which strands come to the fore is likely to vary depending on broader political opportunities.

To some degree, localism reveals the doxa, or the "peace in the feud," that occurs between advocates of mainstream policies and the radical alternatives. The debates largely assume that the central political issue is the degree of participation of the national government in the economy: from very little at the extreme of anarcho-communalists to significantly reduced among neoliberals to moderate and aggressive among social liberals to government ownership among socialists. The terms of the radical and mainstream political debate can be used to inform an analysis of the articulations of localist politics with existing political ideologies, but they can also become a template that fails to reveal the departures from those ideologies. To avoid misinterpretation and to understand localism on its own terms, it is necessary to develop a more succinct vocabulary for its politics.

I suggest that the crucial differences between localist political thought and both radical and mainstream ideologies are the emphasis on the role of small businesses and local nonprofit organizations, the call for independent and local ownership, and the goal of extending that project to locations throughout the world in the form of a global economy based on locally owned independent enterprises, such as in fair trade networks. Local autonomy translates largely into a concern with ownership—that is, the question of who owns the means of production. However, in contrast with both radical and mainstream traditions, localism does not entail framing the ownership issue in terms of more or less public ownership, as occurs in debates over privatization and nationalization. The mainstream political debates focus on more or less government intervention in the economy, and the radical debate pushes either for federal ownership in the socialist tradition or for municipal and communal ownership in the communalist tradition. Localism departs somewhat from the existing political debates by shifting attention from the government-economy relationship to the relationship between multinational corporations and society. At the heart of the concept of local independent ownership is a political project of building an alternative economy distinct from the world of the large, publicly traded corporation. This position has resonances with radical critiques of capitalism, either from a socialist or an anarchist perspective. However, the focus on small-business development through market development and government programs also has resonances with neoliberalism and social liberalism. The strong attention drawn to the shortcomings of a global economy dominated by enormous corporations with little concern for nation-states or for place-based communities, and often with little concern for the environment and hourly workers, represents a kind of politics that seems especially geared toward addressing the problems that have emerged in an era of globalization.

In addition to drawing attention to the large publicly traded corporation as the central unit in need of reform, localism configures the problem of justice in a different way from mainstream approaches. The "peace in the feud" between mainstream and radical debates on justice concerned the problem of social inequality, especially the fates of working-class and poor people. The debate has always been about how to solve the problem of helping those at the bottom of the social ladder, both at home and abroad. The solutions range from neoliberals' emphasis on enterprise development zones and workfare to welfare state liberalism to the redistribution of profits through communal or government ownership. Although the positions are quite different, the overall debate shares an emphasis on justice in the distributive sense of solving social inequality and poverty.

Localist politics broaden the discussion of justice by injecting what might be considered a procedural perspective into the debate. For localist politics a central justice issue is the loss of economic and political sovereignty of place-based communities to global capital, which implements new regimes of governance through control of federal government policies, continental trade agreements, and global trade and financial organizations. By sovereignty I mean nothing more complicated than the traditional understanding of a government's ability to regulate and otherwise control the economics and politics of its territory and population. In a world dominated by multinational corporations, it has become increasingly difficult for local communities, and even large nation-states, to achieve autonomy from the priorities set by global capital. Localism draws attention to an underlying problem that is a precondition for a community or larger political unit to be able to address issues of distributive justice. If the democratic governance of the economy is broken as a result of corporate control of local, state, national, and international governments and governing bodies, then it will be difficult for governments to address significant social and environmental problems. Conversely, a community with high economic sovereignty could be in a better position to address issues of poverty within its boundaries than one that is governed by outside forces. However, the two issues are analytically distinct, and the difference is crucial if one is to understand what localism is about as a form of political thought and action.[6]

To summarize, the political thought that often underlies localist movements in the United States tends to emphasize the problems of the corporatization of the economy and the loss of local sovereignty. It draws attention to the project of building an economy based on economic units other than large corporations, rather than finding solutions that adjust the role of the government in the economy and that address the pervasive growth of within-nation inequality. The problems that preoccupy the ongoing political field of mainstream and radical positions do not disappear, but instead the terms of the debate about the economy, sustainability, and justice are widened. Just as the radical alternatives to mainstream politics opened up a broader set of political issues for consideration and contestation, so localism opens up the debate of mainstream and radical politics and policies to a broader field of issues.

Sometimes described as "new economics" or "alternative economics," the political ideology associated with localist movements can be understood as a historical response to a neoliberal pattern of globalization that has favored the consolidation of major industries, the concentration of wealth, the influence of global economic elites on governments, and the sense that democracy has given way to corporatocracy. From this perspective localism is not necessarily a solution to the global environmental problems. It is likely, as the other contributors to this book suggest, that global environmental problems such as peak oil and climate change will result in a long-term trend of localization. Furthermore, some wings of the localist movement in the United States—especially the organizations affiliated with the Business Alliance for Local Living Economies—forge explicit linkages among local ownership, environmental sustainability, and social equality. However, it would be a mistake to assume that localist politics are necessarily linked to sustainability concerns. Rather, when understood as efforts to build political and economic sovereignty through enhanced local ownership, localist politics can become a defensive political movement anchored in the small-business sector of mom-and-pop shops, local service firms, small farmers, community banks, and independent media.

Although there is a potential for localist politics to become defensive and conservative, to date in the United States there are many examples of localist politics that have become connected with sustainability and justice concerns. Furthermore, as a political movement that challenges a neoliberal vision of the global economy dominated by large, publicly traded corporations, localist politics helps to undermine the power of global corporations that have put profits before people and the planet. When joined with other social movements and a growing countervailing sector of green businesses that have stakes in a sustainability transition, localist movements can contribute to opening political opportunities for a more rapid transition to more sustainable and just forms of organizing society.

Notes

1. For examples of the reporting of the problematic aspects of corporate greening in the business press, see Elgin 2007 and Gunther 2006. See also Sklair 2001 and Weinberg 1998 on the limitations of corporate greening.
2. Meadows, Randers, and Meadows 2004. See also York and Rosa 2003.
3. On disaster capitalism, see Klein 2007.
4. Schumacher 1973; Wood 1984.
5. On the convergences of localism with neoliberalism in the context of rural regions, see DuPuis and Goodman 2005. On the ordinances against franchises and zoning for big-box superstores, see Mitchell 2000, 2006.
6. See also Hinrichs and Allen 2007.

References

DuPuis, Melanie, and David Goodman. 2005. Should we go home to eat? Toward a reflexive politics of localism. *Journal of Rural Studies* 21:359–371.

Elgin, Ben. 2007. Little green lies. *Business Week*, October 29, O45–O52.

Gunther, Marc. 2006. The green machine. *Fortune*, August 7, 42–57.

Hinrichs, C. Clare, and Patricia Allen. 2007. Buying into "buy local": Engagements of United States local food initiatives. In D. Maye, L. Holloway, and M. Kneafsey, eds., *Alternative Food Geographies: Representation and Practice.* Amsterdam: Elsevier.

Klein, Naomi. 2007. *The Shock Doctrine: The Rise of Disaster Capitalism.* New York: Metropolitan Books.

Meadows, Donella, Jørgen Randers, and Dennis Meadows. 2004. *Limits to Growth: The Thirty-Year Update.* White River Junction, VT: Chelsea Green.

Mitchell, Stacy. 2000. *The Home Town Advantage: How to Defend Your Main Street against Chain Stores—and Why It Matters.* Washington, DC: Institute for Local Self-Reliance.

Mitchell, Stacy. 2006. *Big-Box Swindle: The True Cost of Mega-Retailers and the Fight for America's Independent Businesses.* Boston: Beacon Press.

Schumacher, E. F. 1973. *Small Is Beautiful: Economics as if People Mattered.* New York: Harper and Row.

Sklair, Leslie. 2001. *The Transnational Capitalist Class.* New York: Blackwell.

Weinberg, Adam. 1998. Distinguishing among green businesses: Growth, green, and anomie. *Society & Natural Resources* 11 (3): 241–250.

Wood, Barbara. 1984. *E. F. Schumacher: His Life and Thought.* New York: Harper and Row.

York, Richard, and Eugene Rosa. 2003. Key challenges to ecological modernization theory. *Organization & Environment* 16 (3): 273–288.

VI

Tools for Transition

The pessimist complains about the wind; the optimist expects it to change; the realist adjusts the sails.

—William Arthur Ward

Localization is inevitable in some form. To be realistic and positive—that is, for the transition to be peaceful, democratic, just, and ecologically resilient—people everywhere will have to adjust their lives and communities. Given that human societies were once organized locally, such adjustments are clearly possible, maybe even easy once people develop a habit of envisioning positive futures. However, that habit is not easy to acquire, especially when a frenetic, consumerist, fossil fuel–dependent society seems to throw up one roadblock after another (e.g., rising energy prices and health-threatening food amid widespread joblessness). For all this, localizers will need tools, including tools for overcoming barriers to effective change.

While there is much written on tools for, say, living independently on a few acres or producing one's own electricity, there is much less written on tools for addressing humans' psychological and organizational needs in a process like localization.

The next four readings offer such broad tools. Each provides procedural guidelines but may not be explicit about the direction or goal. When combined with earlier readings, however, these tools can help people envision strategies for moving toward positive localization.

Such strategies have three key features. One is the need for localizers to conduct a multitude of small experiments. These must be done quickly, while simultaneously providing support for likely failures. Support for failure is crucial since it can be hard to pursue extreme change when the cost of failure to oneself and one's community is high. In chapter 22, Raymond De Young and Stephen Kaplan propose adaptive muddling, a tool that encourages a family of solutions, not the one best solution. Adaptive muddling encourages small experiments and portions of solutions, both of which remove the need for making the whole transition at once. In chapter 24, Donella Meadows, Jørgen Randers, and Dennis Meadows similarly argue that change agents should never assume they have the entire answer.

A second feature of these tools is the nature of leadership. The adaptive muddling piece introduces distributed leadership, a tool for maintaining direction, continuity, and enthusiasm during what will likely be a long, drawn-out transition. Taking a long historical perspective, Lester W. Milbrath proposes a partnership model in chapter 23. He argues that the dominator model of recent centuries is no more inevitable or natural than any other. In fact, the dominator model has arguably contributed to many of the crises that now compel localization. Perhaps domination was aided and perpetuated by the historically sudden availability of cheap and abundant energy. If true, then the coming energy descent may provide just the nudge societies need to reestablish a partnership ethic.

A third feature is mental change, which in our concluding chapter we call the *downshift moment*. The question here is, what are the conditions under which people pursue new patterns of behavior? Conventional wisdom has it that humans anchor on the status quo, thus making change difficult. But research shows that the issue is not so much a bias toward the status quo as a bias toward *familiarity*. In human cognition, familiarity is the outcome of constructing cognitive maps, a means of making sense of the world. Cognitive maps mirror the strengths and weaknesses of people's current understandings and they enable, or restrict, imagination. Fortunately, cognitive maps can change, sometimes rapidly and radically. Humans are, after all, the quintessential adaptive creature. New information, some experienced directly, some learned from others, and the sharing of information are central to this change.

Meadows, Randers, and Meadows propose steps for adaptive change such as envisioning alternatives, networking to get help and to help others, being upfront about premises and intentions, and questioning the usefulness of certain metaphors. They urge doing all of this with urgency, given the state of affairs, and with humility and in a loving way. These last notions are not common in environmental, let alone scientific, writing. Yet, we editors find in our research that it is long past time to seriously question, then reject, the model of human behavior that presumes people are fundamentally shortsighted, individualistic, and uncaring.

Finally, localizers must acknowledge that, for all the tools available to "adjust the sails," emotions such as dread, denial, and fear are real, especially when the future is hard to imagine and dramatic change is looming. So, besides calling attention to Sharon Astyk's arguments in chapter 16, we flag one additional tool, which De Young and Kaplan elaborate on in chapter 22. This is prefamiliarization, the ability to become thoroughly familiar with a process (such as localization) even though it has only just begun.

22

Adaptive Muddling*

Raymond De Young and Stephen Kaplan

In this book we assert that localization is an inevitable phenomenon rather than a process awaiting someone's initiation. There are no guarantees, however, that localization will move in a desirable direction, toward what we have called positive localization. *To move in this direction, a great many small experiments must be conducted, and quickly. Many need to be far-reaching; some will fail, but all will inform a larger process of societal adaptation to a new biophysical reality.*

What would the planning and decision making look like, especially if the path of localization is uncertain, even treacherous at times, yet urgent? Here Raymond De Young and Stephen Kaplan suggest adaptive muddling. Although muddling may sound less than systematic or rational, they develop the concept in psychological terms, showing that it is very human of us to muddle. In fact, our very success as a species may in part be attributed to muddling's outcomes. Notice that a key feature of muddling is its emphasis on small experiments, not the small steps common in formal decision-making procedures. The difference is subtle but important. Adaptive muddling does not privilege the status quo by investigating only marginal change. Instead, it encourages exploring, and thus prefamiliarizing, for life-changing adaptations. It also offers a way of simultaneously exploring several ideas at once, thus avoiding the sluggishness that plagues conservative, one-solution-at-a-time approaches. And it supports the application of local knowledge to local situations; different people applying different knowledge to the same situation creates a variety of potential solutions. It is such enhanced, rapid, and diverse creativity that localization needs.

*De Young, Raymond, and Stephen Kaplan. 1988. "On Averting the Tragedy of the Commons." *Environmental Management* 12 (3): 273–283. Excerpted and reprinted with permission.

Note also that adaptive muddling contains a stability component that not only reduces the costs of failure for the individuals involved, but also makes unchecked and disorienting change highly improbable. Finally, adaptive muddling inspires leadership that is appropriate for localization—distributed and deliberative, not centralized and controlling—while making rapid experimentation at a scale small enough to be psychologically effective.

The tragedy of the commons (Hardin 1968) has proven a useful concept for framing how we have come to be at the brink of ecological catastrophe. People face a dangerous situation created not by evil outside forces, but by the apparently appropriate and innocent actions of many individuals. This by now widely known paradigm is applicable in its broader sense to a great many environmental problems. The response presented here takes as its starting point two criteria that a solution must meet in order to be successful: compatibility with human nature and compatibility with available natural resources.

While the tragedy of the commons is a distressing event, the proposed authoritarian solution might well be equally disturbing. Nevertheless, the proponents of the authoritarian solution are not Machiavellian. They view the problem to be of such a great magnitude and urgency that people can neither wait for the democratic process to act nor tolerate the resulting compromise solutions. Their approach is viewed as a necessary evil, one that addresses the realities of ecological limits, not political acceptability. Authoritarianism, with its apparent direct and uncompromising approach, is, they argue, the only hope.

Reactance to the Elimination of Choice

While coercive solutions are perceived as overcoming the weaknesses of democratic institutions and human nature, they are not without their own limitations. A problem with solutions that eliminate choice is the undesirable effects they have on individuals (Vargish 1980). The characteristic negative human reaction to strong coercion has been analyzed in the context of psychological reactance theory. Psychological reactance is the motivational state of a person whose freedom has been constrained (Brehm 1966, Brehm and Brehm 1981). It is a response by which people show increased desire for a forbidden alternative or decreased desire for what they feel forced to do. This phenomenon is more than just a disturbing possibility. Reactance effects have been noted in numerous investigations, including the study of legal prohibitions (Mazis 1975) and strongly worded prompts for proenvironmental action (Reich and Robertson 1979). The tendency to react against compulsory changes that involve one without consent appears to be a rather general phenomenon.

Loss of Diversity and the Potential for Grave Error

Nonetheless, some might conclude that the risk of reactance is a cost one must bear. The authoritarian solution is not, after all, a preferred solution; the argument is rather that it is a necessary one. Taking a step that is so decisively against the grain of human nature could, of course, only be justified in terms of its unequivocal effectiveness in the utilization of available resources.

Centralized planning attempts to manage resources, and simultaneously overcome human weaknesses, by applying one pattern to all possible settings. This approach reduces the chance of people "messing things up," but also loses the diversity and resilience so essential to effective resource management. Consider such federal energy conservation efforts as the building-temperature setback program, with one target temperature for all climatic regions and building types. While such federal efforts have been evaluated as generally ineffective, there have been notable successes at the local level, each demonstrating a sensitivity to local conditions (Ridgeway 1979, Stern and Aronson 1984).

A related issue is the ability of authoritarianism to commit a large percentage of available resources to what is judged to be a vital project. While the urgency of the environmental crises would seem to demand such a response, it entails considerable risk. There is the danger of making large-scale resource allocation errors. In fact, the potential for grave errors may be a major risk of the authoritarian approach. As Lindblom (1979) notes,

> authoritarian systems are at least occasionally capable—apparently more often than in democratic systems—of such nonincremental change as the abrupt collectivization of agriculture in the Soviet Union and the Great Leap Forward and the Cultural Revolution in China (as well as the Holocaust and the recent destruction of Cambodia's cities and much of its population).

Thus, authoritarian systems may be less effective at avoiding ecocatastrophe than their proponents believe. What we know to be the limitations of large-scale institutions casts doubt on the appropriateness of an authoritarian approach (Orr and Hill 1978, Lindblom 1979). Such an approach, while consuming valuable resources, may contribute nothing to the problem at hand.

Muddling, in contrast, is compatible with human nature. We seek advice of those people most affected by the decision, check out every step in advance for acceptability, and never venture far from the result of past changes. Muddling is conservative in a functional way: it does not ignore traditional values or beliefs (Ophuls 1977). The outcome of muddling is a practical solution, one that at least makes a marginal contribution to the problem at hand. One of the important contributions that Johnson has made is the fusion of the muddling process and environmental problems (Johnson 1978 and 1985). Johnson has shown that the prevalent human tendency to "make do" can aid us in adopting a frugal way of life without the need for an authoritarian bureaucracy.

Muddling as a Process of Sequential Exploration

It is of considerable interest in solving commons dilemmas to have the ability to explore a variety of potential solutions at the same time. It is our perception that the proponents of the muddling and authoritarian approaches may suffer in this respect. It would appear that they function by pursuing one solution at a time. Certainly they are capable of intelligently discussing alternative courses of action, but the outcome of such discussion usually is the selection of one option for implementation. Furthermore, this selection is more often based on political acceptability than on the feedback of facts or the reports from what one might call field tests. Essentially, an entire society explores a chosen solution. Implementing solutions on a national level without the benefit of pilot testing may strike the reader as risky. The failure of a solution could be costly, disastrous, or both. As an example one need only consider the "payment in kind" program that seemed to be a good solution to the huge US farm product surplus until the solution was implemented nationwide. Clearly, what was perceived to be an effective solution at the conceptual stage proved otherwise. It would seem less than prudent to move quickly from the conceptual stage to full-scale implementation without the benefit of smaller-scale explorations. Nevertheless, it seems to go on all the time.

There are, thus, several reasons why something more than muddling is needed. It is awkwardly slow in practice and tends to pursue only one large-scale exploration at a time. And this exploration involves an often untested solution implemented at a large scale. Without some adjustment, muddling will not be of much help.

Adaptive Muddling

We suggest that it is possible to build upon people's natural tendency to muddle through. It is worth salvaging the powerful advantages of muddling if its tendency toward slowness, compromise, and lack of direction can be overcome. We propose that muddling be improved by introducing three distinct facets of the decision-making process: exploration, stability, and distributed leadership. We are calling this new framework *adaptive muddling*.

A key feature of adaptive muddling is the relationship of scale to function. In the midst of a heated debate between the advantages of large-scale or small-scale solutions, Schumacher (1973) has suggested a middle ground. His concept of "intermediate technology" highlights the importance of finding the appropriate scale for solving a problem, not just the smallest scale. Using this approach, one looks for what each level of scale does best and allocates decision-making responsibility accordingly. In adaptive muddling, explorations are pursued at the smaller scale while stability is provided at a larger scale. As a system, however, adaptive muddling is not scale specific. It is easily as applicable to county-level issues as to national or multinational problem solving.

Exploration

One of the strengths of adaptive muddling is related to the small scale of these explorations. Adaptive muddling allows explorations to be pursued at a smaller scale, making it possible to analyze, design, and implement a variety of alternative solutions simultaneously. Any successful small-scale explorations can be considered for implementation at a larger scale.

Such an approach is tolerant of failure for at no time is the entire resource base jeopardized. And yet adaptive muddling is not risk free. One's success or failure has relevance to the larger context. Failure will be felt by both the exploratory groups and all others who await a solution to the tragedy of the commons. The result is a heightened sense of genuineness and connectedness born of knowing that one is trying to solve an urgent dilemma.

But adaptive muddling offers more than just the ability to explore many potential solutions at once. In contrast to the authoritarian approach, which must work against human ingenuity as it attempts to implement a comprehensive solution, adaptive muddling is based on people being involved in the decision-making process. Adaptive muddling builds on people's desire to participate, to do things that can make a difference in a larger context and that matter in the long run. At a smaller scale, with fewer people involved, it is possible for each individual to comprehend the situation and for the exploration as a whole to be action oriented. People are most capable and effective when dealing with something they comprehend. They also respond well when there is a real opportunity for action (Kaplan and Kaplan 1982).

Stability

To deal successfully with the tragedy of the commons, one must explore alternative solutions. But explorations entail risk and uncertainty: There is, thus, a conflict—exploration is both necessary and dangerous. In such a situation, one benefits from a source of continuing and reliable support. With such support, one is freed of having to develop contingency plans against every possible failure (Cantril 1966).

Providing such stability is the function of a context larger than that of exploration. Stability can occur at the level of the nation, the state, or some other institution. By creating the support structure that permits a variety of explorations to take place, it is possible to experience errors without endangering the entire system. Another advantage of being able to tolerate errors is that it permits the exploration of bold and innovative solutions.

There is, thus, a symbiotic relationship between exploration and stability. A stable and predictable environment provides a safe and secure source of support for exploration. In return, explorations provide tested solutions that can be considered for implementation in the larger context. Thus, explorations, which may be neither tightly interconnected nor centrally managed, are nonetheless tied to the larger context via the stability concept.

Interestingly, this combination of stability and exploration has also been identified as vital to the business world. Those companies that Peters and Waterman (1989) have identified as top performers usually have some sort of mechanism for encouraging small, experimental, "'skunk works,' bands of eight or ten zealots off in the corner. . . ." These groups are usually left with a free hand and are often allowed to operate outside of the corporate chain of command. This lack of orthodoxy does not diminish their perceived value to the corporation. These groups are so valued for their innovative contributions that they are supported (that is, offered job security) even in the event one of the ventures should fail. Peters and Waterman are blunt as to the effectiveness of this approach: "No support systems, no champions. No champions, no innovations."

Another benefit of permitting many separate explorations to take place is the possibility of discovering solutions that address local conditions (Peters 1985, Runge 1985). Rather than a single experiment, adaptive muddling supports simultaneous explorations. While decisions are thus repeated many times over, the likelihood is increased that each exploration will include more of the unique facets of the problem in its setting. In the process, a diversity of solutions emerges. In this manner, regional differences in environmental conditions or resource availability are addressed directly, not through the ad hoc modification of a master plan.

Distributed Leadership

An unintended consequence of a coercive solution is, as we have seen, that people are inclined to act in opposition to what they are told to do. This will be true whether the source of the solution is from an authoritarian bureaucracy as in authoritarianism or from a panel of experts as in most democratic decision-making processes. Adaptive muddling, by contrast, benefits from the insight and talent of the entire nation, experts and citizens. A major way in which people feel needed and can contribute to the problem-solving process is through distributed leadership.

Leadership has little or no explicit role in ordinary muddling; by contrast, in adaptive muddling, leadership is central. At the same time, leadership is not seen as residing in one or even a few individuals. The leadership appropriate to adaptive muddling depends upon contributions of many kinds, coming from people representing diverse skills, abilities, and interests. The effectiveness of distributed leadership depends upon its broad base of contribution and the diversity of its contributors (Wildavsky 1964). This can perhaps best be appreciated in the context of adaptive change. The clear message of the commons dilemma analysis is that human patterns must change, and change far faster than can be accommodated by biological modification. The burden, then, is necessarily on cultural rather than genetic evolution. Convincing a culture to adopt new patterns is admittedly a difficult task. Several factors, however, can facilitate the process. The provision of these facilitating factors is the role of distributed leadership. There are three distinct functions here, which might appropriately be labeled *vision*, *process*, and *themes for exploration*.

The first function of distributed leadership is *vision*—the creating of an understanding of our situation, the possibility of a solution, and the challenge it offers to everyone. The urgency of the tragedy of the commons must be understood by all. There must be a shared understanding that this dilemma must be faced, not deliberately avoided. The very nature of the dilemma causes some individuals to feel helpless. Leadership must show that not only can a decent lifestyle be salvaged, but that there is the possibility of improving our quality of life in the process.

Another function of leadership is to create and support a *process* for meeting the challenge of the tragedy of the commons. Leadership must clarify the role of exploration and stability. It is necessary to understand that the failure of explorations is an acceptable, and even necessary, outcome of the process.

And finally, it is the role of leadership to identify and nominate *themes for exploration*. It is difficult to generate ideas that are worth pursuing. Themes that are worth exploring must fit a certain profile: they must deal with the problem and be compatible with human nature and available resources. Leadership in this context is more than just power or charisma. It involves insight that appreciates both the nature of the problem and of the people who must ultimately deal with it.

Adaptive Muddling and the Problem of Small Steps

A key issue as far as adaptive muddling is concerned is how the potentially fatal "small step" bias of conventional muddling can be corrected. As long as only modest steps are taken, the solution to problems will necessarily be slow and frequently too slow. Adaptive muddling deals with this difficulty on two fronts.

One source of reactance to new ideas and new alternatives is the fear of change. While this may seem to be a rather foolish and small-minded fear, in recent times it has all too often been justified. Whether justified or not, however, such a fear can readily block innovation before a new possibility has even had a fair hearing. Here the stability aspect of adaptive muddling plays a central role. It provides assurance that unchecked and disorienting change is unlikely to occur. The stability-preserving feature of adaptive muddling will tend to block the widespread implementation of untested solutions.

A second factor is even more central. Much of what is considered a bias toward the status quo is in fact a bias toward the familiar. Although this may sound as if it is a rather small, perhaps even academic, distinction, it is in fact a crucial issue and one with major implications. A status quo bias means that little will change, and what does change will change only very slowly. A familiarity bias, by contrast, means that the size of the step is limited not by where people are, but by what they know. The key issue, in other words, is the sort of knowledge and experience that people have, not merely what their current circumstances happen to be. Fortunately, people can have knowledge and experience that extend far beyond the comfortable outlines of the status quo.

It should perhaps be emphasized that a familiarity bias is not an expression of a foolish conservatism, but is in general a sound and reasonable practice. To move significantly beyond what one is familiar with is an invitation to disaster. Being on unfamiliar ground, one not only does not know what to do if something goes wrong; one may not even know what "wrong" looks like. Even the familiar is susceptible to surprise and uncertainty; managing these hazards outside of one's comprehension and confidence is a risky proposition indeed.

Since people are conceptual animals, what they can become familiar with is, fortunately, not limited to what they have experienced in a direct and literal sense. People can acquire familiarity through the written word, through artistic creations, and through rough simulations of various kinds (for instance, plays, TV, and computers). Here both leadership and exploration have key roles. Leadership can provide the imagery and the richness of context necessary to allow people to build models of the not yet present. Exploration can provide concrete alternatives that deserve attention and thought.

Conclusion

The authoritarian solution is characterized by the limited role offered individual citizens. There is room for their approval of the selected solution, but no opportunity for them to influence the choice. Even in democracies, the situation is often far from ideal. Williams (1986) describes citizen participation as "citizens (usually sitting passively in chairs) posing questions and comments to experts (if male, usually wearing a necktie). Two messages are clear: Citizens only have input as outsiders into a process controlled by professionals; and citizen involvement (as both obligation and privilege) is limited to planning." The success of adaptive muddling, by contrast, is both sensitive to human concerns and relies on the talent of its citizens. Again, Williams (1986) suggests that citizens "have more than just input into a process; they also shape and define the project through their own actions."

A similar contrast exists for the type of outcome desired by each decision-making process. The authoritarian approach seeks a comprehensive and unidimensional solution. Conventional muddling, with its reliance on due process, is neutral on the desired outcome. If a solution proceeds all the way through the tortuous procedure of due process, it will be acceptable. In contrast, the type of solution to emerge from adaptive muddling is largely independent of the problem-solving process. The solutions developed will be mainly influenced by the particular people involved in the adaptive muddling process and, via this direct involvement, a substantial increase in the overall commitment to the solution will develop.

The advantages of adaptive muddling over Hardin's mutual coercion mutually agreed upon are of a similar nature. In contrast to the perceived bleakness of mutual coercion, adaptive muddling offers a way for people to be inspired by challenge. The necessary solutions, far from being obvious, will require creativity and innovation. Rather than an enforced uniformity, there is likely to be considerable diversity in the patterns of solutions adopted by different individuals. By engaging the imagination and preserving choice, adaptive muddling gains a substantial benefit both politically and psychologically.

There is growing evidence that behavior is strongly affected by the model of the world one holds. What is lacking in the present situation is a model that relates what experimenting does occur to the decision-making process. Thus, implementation requires, at its roots, a change in the shared model of the governmental process. We have attempted to sketch out a blueprint for this needed model. Its major components can be described concisely. It requires acknowledging a problem (that is, environmental limitations) that tends to be denied. It requires a clear policy to the effect that (a) outcomes matter, (b) these outcomes cannot be known without exploration, (c) this exploration is best done at a small scale, and (d) in order to find solutions in a timely fashion, many such experiments must go on simultaneously.

Thus, the primary ingredients already exist. The talent exists. The problems exist. The history of muddling exists. What is required is the adoption of a new way of thinking about how these components can be brought together into a larger whole that is challenging, inspiring, and effective.

References

Brehm, J. W. 1966. *A theory of psychological reactance.* Academic Press, New York.

Brehm, S., and J. W. Brehm. 1981. *Psychological reactance: a theory of freedom and control.* Academic Press, New York.

Cantril, H. 1966. *The pattern of human concerns.* Rutgers University Press, New Brunswick, New Jersey.

Hardin, G. J. 1968. "The tragedy of the commons." *Science* 162:1243–1248.

Johnson, W. 1978. *Muddling toward frugality.* Shambhala, Boulder, Colorado.

Johnson, W. 1985. *The future is not what it used to be: returning to traditional values in an age of scarcity.* Dodd, Mead and Company, New York.

Kaplan, S., and R. Kaplan. 1982. *Cognition and environment: functioning in an uncertain world.* Praeger, New York.

Lindblom, C. E. 1959. "The science of muddling through." *Public Administration Review* 19:79–99.

Lindblom, C. E. 1979. "Still muddling, not yet through." *Public Administration Review* 39:517–526.

Mazis, M. R. 1975. "Antipollution measures and psychological reactance theory: a field experiment." *Journal of Personality and Social Psychology* 31: 654–660.

Ophuls, W. 1977. *Ecology and the politics of scarcity.* Freeman, San Francisco.

Orr, D., and S. Hill. 1978. "Leviathan, the open society, and the crisis of ecology." *Western Political Quarterly* 31:457–469.

Peters, P. E. 1985. Concluding statement. In *Proceedings of the conference on common property resource management*, 617–621. National Academy Press, Washington, DC.

Peters, T. J., and R. H. Waterman. 1982. *In search of excellence: lessons from America's best-run companies.* Harper and Row, New York.

Reich, J. W., and J. L. Robertson. 1979. "Reactance and norm appeal in antilittering messages." *Journal of Applied Social Psychology* 9:91–101.

Ridgeway, J. 1979. *Energy-efficient community planning: a guide to saving energy and producing power at the local level.* JG Press, Emmaus, Pennsylvania.

Runge, C. F. 1985. "Common property and collective action in economic development." In *Proceedings of the conference on common property resource management*, 31–60. National Academy Press, Washington, DC.

Schumacher, E. F. 1973. *Small is beautiful.* Harper Torchbooks, New York.

Stern, P. C. 1978. "When do people act to maintain common resources? A reformulated psychological question for our times." *International Journal of Psychology* 13:149–158.

Vargish, T. 1980. "Why the person sitting next to you hates limits to growth." *Technological Forecasting and Social Change* 16:179–189.

Wildavsky, A. B. 1964. *Leadership in a small town.* Bedminster, Totowa, New Jersey.

Williams, H.S. 1986. "Barn raisings: an old approach helps small towns today." *Small Town* (May-June):14–15 and 18–19.

23

Promoting a Partnership Society*

Lester W. Milbrath

Political scientist Lester W. Milbrath directs our attention to the drivers of localization (see part I) by showing how dominator societies *have organizational and behavioral norms that create conditions for societal collapse. He brings to light modern, often hidden presumptions about the best way to maintain social order. But he does more than critique. He also presents the* partnership society, *a hopeful and positive alternative to dominator societies, and suggests that such societies are plausible because they once existed, often for long periods of time.*

Milbrath concludes with a response to those who say that trying to repair the world is futile because human behavior is such that good works are always undone. Localizers will find his talking points useful when they encounter opponents who believe humans are fundamentally flawed, unable to escape their unreasonable, selfish, hypercompetitive, and warlike selves. Most heartening in Milbrath's argument is the claim that today's problems are not so large, nor the current social and political systems so powerful, that ordinary people are helpless to do good.

Humans, among all species, can invent their pattern of life. But if that is true, why do we seem unable to shape our destiny? Why are wars so frequent? Why do men dominate women? Why is there so much suffering and privation? Why do we wreak havoc on our ecosystem? Why do we fear that the next war will annihilate most life on the planet?

Nearly all societies are dominator societies in which some people rule other people. All of us grew up in a dominator society. Many take

*Milbrath, Lester W. 1989. "Transforming the Dominator Society." In *Envisioning a Sustainable Society: Learning Our Way Out*, 39–57. Albany: State University of New York Press. Excerpted and reprinted with permission.

domination to be characteristic of all societies. Is that true? Is it inherent in human nature for some people to dominate others? Men are taught to be competitive and aggressive; if they do not do this, others perceive them as not being *real* men. A window sticker on a van parked in my university's parking lot proclaims: "God, Guns, Guts Made America Great; Let's Keep All Three." Is that true? Most women have experienced male domination; they are even socialized to admire and reward it. Capitalist doctrine proclaims the competitive society as the best. Is that true? People in the United States are so competitively oriented that a football coach has been eulogized for proclaiming, "Winning isn't everything, it's the *only* thing."

Many social problems can be traced to the competitive struggle for money, power, prestige, and control. Why do we join this struggle and why is it so difficult to avoid it? How does this struggle interfere with our efforts to develop a society that can live in a long-run harmonious relationship with nature and in which people can live in peace with each other? How did we get into this predicament? Perhaps we can gain insight by taking another look at history.

A New Perspective on Human History

. . . The historical perspective sketched here is quite different from the one I was taught as a schoolboy. My new perspective has been stimulated by three recent books: Schmookler's *The Parable of the Tribes* (1984), Eisler's *The Chalice and the Blade* (1987), and Margulis and Sagan's *Microcosmos* (1986). These reinterpretations of history reflect very recent scholarship; none were written by historians. Their new data come mainly from archaeological, anthropological, fossil, and geological records. They show us that traditional history, which is derived primarily from written records, can lead us to misinterpret the essence of human nature and human society.

Eisler (1987) summarizes considerable archaeological evidence showing that relations between people in some of the cultures created by . . . [the] capability for learning may have been very different from those we know today. She characterizes them as "partnership" societies in contrast to the "dominator" societies of today. Because they were not patriarchal, we should not assume they were matriarchal; rather, the archaeological evidence suggests that neither sex dominated the other—they lived together in partnership. . . .

Both Schmookler and Eisler dwell on the way the civilization that emphasizes domination has pervaded all aspects of life and spread to the remotest parts of the globe. Both agree that the struggle for power and domination is not inevitable; other successful societies existed where people felt equal and did not have to struggle for position. Why did partnership societies give way so completely to the dominating conquerors? Schmookler's "parable of the tribes" provides an explanation.

Schmookler's parable deals with tribes, or societies, that do not have a governing arrangement that regulates their relationships; it is analogous to anarchy among individuals. As human numbers grew and tribes enlarged, societies confronted each other. Each society in such a confrontation faced an unpleasant choice. If it willingly stopped its growth so as not to infringe on its neighbors, it could foresee that death would catch up and overtake it. If it continued to expand, it committed aggression.

As civilization developed, humans confronted a situation in which the play of power was uncontrollable: "In an anarchic situation like that, no one can choose that the struggle for power shall cease. But there is one more element in the picture: *no one is free to choose peace, but anyone can impose upon all the necessity for power*. This is the lesson of the parable of the tribes" (Schmookler, 1984 p. 21; emphasis in the original). Evolution under civilization developed a new selection principle: POWER. The evolutionary principle discovered by Darwin asserted that species survived because they found their niche. (Darwin's principle is often misinterpreted as "survival of the fittest" and is used to justify aggressive domination of others.) However, in the evolution of civilization, those cultures survived that best deployed power. The selection for power applies mainly to the struggle among humans, although that struggle often resulted in great harm to other species as well.

Schmookler spends most of his book showing the numerous ways that the struggle for power has enslaved us all and permeates every part of our life. He sees power as a contaminant, a disease, which once introduced will gradually yet inexorably become universal in the system of competing societies. Power dominates because it can prevail; but what prevails may not be best for people and other creatures. The continuous selection for power closed off many humane cultural options that people might otherwise have preferred. Power came to rule human destiny (Schmookler, 1984, 22–23).

I have room here only to summarize the many consequences for society, the ecosystem, and the quality of individual lives that flow from the selection for power, most of which is drawn from Schmookler:

1. The warlike eliminate the pacifistic and content. A tribe (society) that is confronted by an aggressive power-maximizing neighbor has only four options:

   a. It may suffer destruction; in a struggle for power the surviving society will be the one that employs power most effectively.

   b. It may be absorbed by the aggressive power and become transformed into a power-maximizing society.

   c. It may escape the compelling pressures of the intersocietal system by withdrawing beyond the reach of other societies. Only in the least accessible regions of the planet have the equalitarian and peaceful societies been able to survive into our time. All other societies were drawn into the power contest, or were eliminated by it.

   d. If it chooses to defend itself against the aggressor, the peace-seeking society becomes the imitator of the power-maximizing society: "The tyranny of power is such that even self-defense becomes a kind of surrender. Not to resist is to be transformed at the hands of the mighty. To resist requires that one transform oneself into their likeness. Either way, free human choice is prevented. *All ways but the ways of power are blocked*" (Schmookler, 1984, 54; emphasis in the original). A society may also choose to escalate its power, thus starting arms races; this might be thought of as a fifth option. It constitutes total capitulation to the power struggle.

2. Selection for power favors those who exploit nature and discards those who revere it. A society that exploits its resources quickly accrues more power than a similarly based society that husbands its resources and protects its biosphere. The resource-exploitative society may then overpower the more nature-protective society and seize its resources for additional quick exploitation. The drive for power produces an ever-widening gap away from the natural: "Man's power over Nature means the power of some men over other men with Nature as its instrument" (C. S. Lewis, 1946, 178).

3. Larger populations and greater land areas contribute to power; therefore, there will be a tendency toward larger societies: "Size confers power and power facilitates expansion" (Schmookler, 1984, 82).

4. There will be a tendency toward complexity, specialization, and efficiency because they enhance power. Centralization is needed in order to

manage this complexity effectively. Central control enforces unity of purpose, directs coordination of the parts, and induces the parts to sacrifice. The division of labor in a complex system enhances power but it detracts from the wholeness of humans. The demand for power creates a need for drudges.

5. There is a need to be constantly on the alert. We have become an adrenalin society, always organized and oriented toward the requirements of maintaining power and being ever alert to ward off threats from other societies. Even in the times between wars there is no peace.

6. Intersocietal competition has produced in modern times a single global competitive system that is bringing all cultures toward convergence. Increasingly there is but one way into the future, the technological way. We compete in technological development because that is the key to power. Similarly, we are urged to work hard, to save, to produce capital, and strive to become wealthy because this mode of living is superior for accumulating power. This thrust for power builds an ethic that overvalues productivity and undervalues conservation and protection of nature.

7. Reason especially is brought into service to enhance power. Reason, via science, creates economic and military power. The intellectual sphere, in service to power, is permeated with the paradoxical teaching of the value of "value neutrality." Knowledge and technology created by a "value-free science" become the captive of the power-maximizing system: "People who cannot experience their own ultimate purposes provide a vacuum to be filled by the purposes inherent in their systems. . . . The way is cleared for the purposes of the systems to be adopted by the people as their own purposes, rather than vice versa. Technology emerges as the trend of the modern world: rule of the tool" (Schmookler, 1984, 202–203).

8. In a synergistic system the interaction of the parts contributes to the good functioning of the whole; but the ceaseless struggle for power in modem civilization creates a societal environment that is unsynergistic:

a. Conflict gains an ever-increasing sway even though humans do not wish it.

b. The competitive intersocietal system produces a minus-sum game in which all the parties lose.

c. Power is corrupting; it does not serve an essential life sustaining function for the collectivity as a whole.

Schmookler closes his book by offering only a hint of the steps that might be taken to extricate humanity from its horrible predicament. He does look to government as the only social device available for possibly extricating ourselves: "Anarchy enthrones force and only government can place anything else on the throne." But how do we rise above the race for power in most of our nations? How can we find peace in the anarchy of our nation-state system? No one, currently, can answer those questions.

Schmookler's analysis uncovers the deep and fundamental flaws of modern civilization. We can never live in peace, love, and justice in a clean and nourishing environment, we can never have a high quality of life, as long as we retain our power-maximizing, competitive society. We cannot solve the problems of war and peace, or protect the environment, by developing more and better technology. Societal tinkering with new laws or new policies may ameliorate some problems, but will not deflect us from the abyss *as long as the present power-maximizing system is retained*. The only choice that has any hope of saving our species and providing humans with a reasonable quality of life is to transform the dominator society—we must redesign the most fundamental relationships in our civilization.

Eisler dwells much more than Schmookler on the tendency of societies since the dawn of civilization to develop dominator/submissive relationships, especially a patriarchal relationship within families and between the sexes. The societies that conquered the partnership societies emphasized the power that takes away more than the power that gives life; their patriarchal system forcibly asserted male dominion over women, even to deciding their life and death. . . .

The new dominators were not content until they completely transformed the way people perceived reality. As this thrust proceeded, the power of the blade became idealized. Both men and women were taught to equate true masculinity with violence and domination and to regard men who did not conform to this ideal as weak or effeminate. Even today in the United States the most desired characteristic of a President is that he should be a *strong leader*. For those brought up in this system (which includes most of us), believing there is any other way to structure human society is difficult. . . .

Recognizing That the Dominator System Is Maladaptive

The male dominator system seems supremely effective; it has driven out partnership societies; its central objective, power, has guided cultural evolution; it has penetrated to the far reaches of the globe; it has shown extreme resiliency in defending against the occasional resurgence of feminism. However, it has not recognized that it has created a civilization that cannot be sustained, and it has not been able to prevent learning that does not please it. . . . In short, the ways of power and dominance have now accrued so much destructive capability that we can destroy most life on the planet. Even if we avoid that catastrophe, but continue on our present course, we will devastate the ecosphere on which we all depend for life.

Rather than proceed blithely on to breakdown, we must strive in numerous ways to break through to a new, deeper, and more sensitive understanding of our predicament and of what it requires of us. A rising awareness of our predicament, and of patriarchy's role in creating the predicament, is a necessary first step in overturning the dominator society. The strong feminist movement of the last two decades has accomplished a great deal in raising awareness of the subtle and pervasive effects of patriarchy throughout society. That awareness is still confined primarily to women, however, and now needs to be spread to men. Many men will be receptive, especially if they perceive that a change in beliefs and values is essential for saving our ecosystem and society. We can be sure, however, that many other men will feel threatened by partnership ideas and values and will vigorously oppose the change.

The struggle is inevitable, but our aim is to transform conflict rather than to suppress it or explode it into violence. Unlike the struggle of 4,000 years ago when partnership was twisted into patriarchy, this one can probably be won. What has changed? *First*, the world is now very different; the threats of nuclear war, overpopulation, resource depletion, ecosystem devastation, climate change, famine, and disease are so real and imminent that billions of people are aware that drastic changes are needed. System collapse has a remarkable way of freeing one's mind from old conceptions. Nature will not leave us content with our old ways.

Second, our society has learned to abandon feudalism, slavery, and colonialism. We no longer permit men to brutalize women and children. Brute force cannot be used to ensure continued male dominance over women.

Third, in more and more societies decisions are made by votes instead of guns. In the last two decades, dictatorships have more or less peacefully given way to popularly elected governments in Spain, Portugal, Argentina, The Philippines, Brazil, and South Korea. Attempts by superpowers to impose their ideologies by military might in Vietnam and Afghanistan have failed. . . .

Fourth, in nearly all countries that permit elections, women have an equal chance with men to vote and they are learning how to use their vote. The most recent turnout figures in the United States show that women are more likely than men to vote. . . . My own studies show that the pro-environmental protection stance on most policy issues is more likely to be supported by women than men; on average, 15 percent more women than men would support environmental protection.

"In contrast to men, who are generally socialized to pursue their own ends, even at the expense of others, women are socialized to see themselves primarily as responsible for the welfare of others, even at the expense of their own well-being" (Eisler, 1987, 189). In my judgment, women have a much better chance than men of saving the planet. We men who perceive the crucial importance of female participation should do all we can to support their efforts. All of us, both sexes, have both male and female feelings and needs but each sex feels unfree to express both sides of its personality. Liberation from the dominator belief system and social structure will be freeing to both men and women.

Speaking Truth for Learning

Occasionally, we all hear aphorisms that purportedly express truth about human nature and human society. For example, I have often heard these phrases: "We have always had wars and we always will have"; "People are naturally selfish and competitive." Probably the personal philosophy and worldview of many people is a mere collection of such aphorisms. Aphorisms have a surface truth quality that may not survive a close analysis, but they persist as part of the myth structure of society because they are not clearly falsifiable. They also serve as excuses for not acting to make a better society.

The analysis presented above provides a basis for falsifying some of these aphorisms, and it presents an opportunity for each of us to make a contribution to social learning by challenging their presumed validity when they are brought up in conversation. Let us examine the validity of a few of the more common aphorisms:

1. *We have always had wars and we always will have.* A variation on this is, *Human nature being what it is, we will always have war and conflict.* These statements erroneously assume that wars have been present throughout human history and that being warlike is rooted in human nature. Eisler's evidence discloses that civilizations existed for thousands of years with few wars. Schmookler helps us to see that war

is more rooted in the structure of the dominator civilization than in the essence of being human. Humans have a strong desire and need for peace that far outweighs their urgings toward war. If we desire to reduce the probability of war, we should try to change our beliefs and our social structure so that humans are not pressed into going to war.

2. *People are naturally aggressive and competitive.* This statement also implies that aggression is rooted in our biological makeup. Is this so? As much evidence is found that people are naturally loving and cooperative as that they are aggressive and competitive. Social structures and sex roles put people into positions where the only posture that makes sense is to be aggressive and competitive. The behavior demanded of people in business competition is a good example of structurally fostered aggression; failing to be aggressive results in being eliminated from the competition. But, we should keep in mind that much of life does not require aggression and that we can be successful in achieving what we want by acting cooperatively and lovingly. As a matter of fact, recent research shows that cooperation is more likely to be successful over a long time than aggressive competition (for example, Axelrod, 1984). Some societies emphasize competition on the presumption that it will lead to a better society, while other societies emphasize cooperation for the same reason. The one emphasis is just as compatible with human nature as the other. It is difficult to imagine any society being totally competitive or totally cooperative, but we can play some role in tilting the emphasis toward cooperation.

3. *Men will always dominate women.* Again Eisler's and Schmookler's reviews of history show that partnership societies existed for thousands of years before male domination became the norm. Continued domination of men over women mainly results from a belief on the part of both men and women that domination is justified. Those beliefs and their supporting social structure can be changed.

4. *People first look out for themselves.* The implication is that people are basically selfish and will always put their own interests first; that asking people to subordinate their personal interests to those of the group or community is contrary to human nature. The economic theory of modern competitive capitalism is based on this premise. While it holds for much economic behavior, this premise certainly does not explain the following kinds of human behavior: parents sacrificing for their children; citizens giving time and money to group or community endeavors; taking the trouble to vote in elections when you know your vote cannot affect the outcome; working to restore and maintain a clean environment when

everyone benefits from your efforts, even those who do nothing; soldiers sacrificing their lives in war; terrorists gladly sacrificing their lives for their cause. Social scientists have conducted extensive experiments based on the "prisoner's dilemma" game, which assumes that people will act in their personal self-interest. These experiments have repeatedly shown that people will sacrifice their personal interest for the good of the group, especially if the opportunity is provided for knowing and understanding what the group interest is (for an example see Van de Kragt, Orbell, and Dawes, 1983).

5. *The System is so big, powerful, and unyielding that there is no use trying to change it.* This belief is widely held and based on considerable evidence; it also is supported by Schmookler's analysis of power. Yet, as indicated above, history provides many examples of social systems that changed: feudalism, slavery, and colonialism have virtually disappeared; in recent decades, several societies peacefully moved from dictatorships to democracy. Much of this book is devoted to examining the possibility for system change and to suggesting ways for ordinary people to help further that change. Finally, our analysis shows that we have no choice but to change. Facing that prospect, it is wiser to believe that ordinary people can help bring about change than to deny cynically that change is possible.

References

Axelrod, Robert. 1984. *The Evolution of Cooperation.* New York: Basic Books.

Eisler, Riane. 1987. *The Chalice and the Blade: Our History, Our Future.* San Francisco: Harper & Row.

Lewis, Clive Staples. 1946. *That Hideous Strength.* New York: Macmillan.

Margulis, Lynn, and Dorion Sagan. 1986. *Microcosmos: Four Billion Years of Evolution from Our Microbial Ancestors.* New York: Summit Books.

Schmookler, Andrew B. 1984. *The Parable of the Tribes: The Problem of Power in Social Evolution.* Berkley: University of California Press.

Van de Kragt, Alphons, John M. Orbell, and Robyn M. Dawes. 1983. The Minimal Contributing Set as a Solution to Public Goods Problems. *American Political Science Review* 77(1):112–22.

24

Tools for the Transition*

Donella Meadows, Jørgen Randers, and Dennis Meadows

In collaboration with her colleagues, Donella Meadows, systems analyst and self-described farmer, was an early pioneer confronting the contradiction of endless material growth on a finite planet. She and her colleagues took much heat and ridicule for their work. Now, as natural and social systems are stressed to the point of failure—think fisheries and freshwater supplies, housing and finance—and science predicts the continuation of this pattern, Meadows and her colleagues are being proved right, even if the specifics vary.

In their chapter, notice how they draw lessons, not for systems analysts—their colleagues—but for system change, especially personal and societal change in the face of tightening biophysical constraints. Although the authors frame their argument in terms of sustainability, notice how locality is essential both ecologically and socially. And while they argue that the "sustainability revolution" cannot be "dictated," thoughtful people can positively influence the transition's direction, first by changing key properties of systems including relevant information, rules and goals, and material consumption, and then by conducting lots of small experiments.

We have been writing about, talking about, and working toward sustainability for over three decades now. We have had the privilege of knowing thousands of colleagues in every part of the world who work in their own ways, with their own talents, in their own societies toward a sustainable society. When we act at the official, institutional level and

*Meadows, Donella, Jørgen Randers, and Dennis Meadows. 2004. "Tools for the Transition to Sustainability," in *Limits to Growth: The 30-Year Update*, 265–284. White River Junction, VT: Chelsea Green Publishing Company. Excerpted and reprinted with permission.

when we listen to political leaders, we often feel frustrated. When we work with individuals, we usually feel encouraged.

Everywhere we find folks who care about the earth, about other people, and about the welfare of their children and grandchildren. They recognize the human misery and the environmental degradation around them, and they question whether policies that promote more growth along the same old lines can make things better. Many of them have a feeling, often hard for them to articulate, that the world is headed in the wrong direction and that preventing disaster will require some big changes. They are willing to work for those changes, if only they could believe their efforts would make a positive difference. They ask: *What can I do? What can governments do? What can corporations do? What can schools, religions, media do? What can citizens, producers, consumers, parents do?*

Experiments guided by those questions are more important than any specific answers, though answers abound. There are "50 things you can do to save the planet." Buy an energy-efficient car, for one. Recycle your bottles and cans, vote knowledgeably in elections—if you are among those people in the world blessed with cars, bottles, cans, or elections. There are also not-so-simple things to do: Work out your own frugally elegant lifestyle, have at most two children, argue for higher prices on fossil energy (to encourage energy efficiency and stimulate development of renewable energy), work with love and partnership to help one family lift itself out of poverty, find your own "right livelihood," care well for one piece of land; do whatever you can to oppose systems that oppress people or abuse the earth, run for election yourself.

All these actions will help. And, of course, they are not enough. Sustainability and sufficiency and equity require structural change; they require a revolution, not in the political sense, like the French Revolution, but in the much more profound sense of the Agricultural or Industrial Revolutions. Recycling is important, but by itself it will not bring about a revolution.

What will? In search of an answer, we have found it helpful to try to understand the first two great revolutions in human culture, insofar as historians can reconstruct them.

The First Two Revolutions: Agriculture and Industry

About 10,000 years ago, the human population, after millennia of evolution, had reached the huge (for the time) number of about ten million. These people lived as nomadic hunter-gatherers, but in some regions their numbers had begun to overwhelm the once-abundant plants and game. To adapt to the problem of disappearing wild resources they did two things. Some of them intensified their migratory lifestyle. They moved out of their ancestral homes in Africa and the Middle East and populated other areas of the game-rich world.

Others started domesticating animals, cultivating plants, and staying in one place. That was a totally new idea. Simply by staying put, the protofarmers altered the face of the planet, the thoughts of humankind, and the shape of society in ways they could never have foreseen.

For the first time it made sense to own land. People who didn't have to carry all of their possessions on their backs could accumulate things, and some could accumulate more than others. The ideas of wealth, status, inheritance, trade, money, and power were born. Some people could live on excess food produced by others. They could become full-time toolmakers, musicians, scribes, priests, soldiers, athletes, or kings. Thus arose, for better or worse, guilds, orchestras, libraries, temples, armies, competitive games, dynasties, and cities.

As its inheritors, we think of the Agricultural Revolution as a great step forward. At the time it was probably a mixed blessing. Many anthropologists think that agriculture was not a better way of life, but a necessary one to accommodate increasing populations. Settled farmers got more food from a hectare than hunter-gatherers did, but the food was of lower nutritional quality and less variety, and it required much more work to produce. Farmers became vulnerable in ways nomads never were to weather, disease, pests, invasion by outsiders, and oppression from their emerging ruling classes. People who did not move away from their own wastes experienced humankind's first chronic pollution.

Nevertheless, agriculture was a successful response to wildlife scarcity. It permitted yet more population growth, which added up over centuries to an enormous increase, from 10 million to 800 million people by 1750. The larger population created new scarcities, especially in land and energy. Another revolution was necessary.

The Industrial Revolution began in England with the substitution of abundant coal for vanishing trees. The use of coal raised practical problems of earth-moving, mine construction, water pumping, transport, and controlled combustion. These problems were solved relatively quickly, resulting in concentrations of labor around mines and mills. The process elevated technology and commerce to a prominent position in human society—above religion and ethics.

Again everything changed in ways that no one could have imagined. Machines, not land, became the central means of production. Feudalism gave way to capitalism and to capitalism's dissenting offshoot, communism. Roads, railroads, factories, and smokestacks appeared on the landscape. Cities swelled. Again the change was a mixed blessing. Factory labor was even harder and more demeaning than farm labor. The air and waters near the new factories turned unspeakably filthy. The standard of living for most of the industrial workforce was far below that of a farmer. But farmland was not available; work in a factory was.

It is hard for people alive today to appreciate how profoundly the Industrial Revolution changed human thought, because that thought still shapes our perceptions. In 1988 historian Donald Worster described the philosophical impact of industrialism perhaps as well as any of its inheritors and practitioners can:

> The capitalists . . . promised that, through the technological domination of the earth, they could deliver a more fair, rational, efficient and productive life for everyone . . . . Their method was simply to free individual enterprise from the bonds of traditional hierarchy and community, whether the bondage derived from other humans or the earth . . . that meant teaching everyone to treat the earth, as well as each other, with a frank, energetic, self-assertiveness. . . . People must . . . think constantly in terms of making money. They must regard everything around them—the land, its natural resources, their own labor—as potential commodities that might fetch a profit in the market. They must demand the right to produce, buy, and sell those commodities without outside regulation or interference. . . . As wants multiplied, as markets grew more and more far-flung, the bond between humans and the rest of nature was reduced to the barest instrumentalism. (Worster, 1988)

That bare instrumentalism led to incredible productivity and a world that now supports, at varying levels of sufficiency, 6,000 million people—more than 600 times the population existing before the Agricultural Revolution. Far-flung markets and swelling demands drive environmental exploitation from the poles to the tropics, from the mountaintops to the ocean depths. The success of the Industrial Revolution, like the previous successes of hunting-gathering and of agriculture, eventually created its own scarcity, not only of game, not only of land, not only of fuels and metals, but of the total carrying capacity of the global environment. Humankind's ecological footprint had once more exceeded what was sustainable. Success created the necessity for another revolution.

The Next Revolution: Sustainability

It is as impossible now for anyone to describe the world that could evolve from a sustainability revolution as it would have been for the farmers of

6000 BC to foresee the corn and soybean fields of modern Iowa, or for an English coal miner of 1750 AD to imagine an automated Toyota assembly line. Like the other great revolutions, though, the coming sustainability revolution will also change the face of the land and the foundations of human identities, institutions, and cultures. Like the previous revolutions, it will take centuries to unfold fully—though it is already underway.

Of course no one knows how to bring about such a revolution. There is not a checklist: "To accomplish a global paradigm shift, follow these twenty steps." Like the great revolutions that came before, this one can't be planned or dictated. It won't follow a list of fiats from a government or from computer modelers. The sustainability revolution will be organic. It will arise from the visions, insights, experiments, and actions of billions of people. The burden of making it happen is not on the shoulders of any one person or group. No one will get the credit, but everyone can contribute.

Our systems training and our own work in the world have affirmed for us two properties of complex systems germane to the sort of profound revolution we are discussing here.

First, information is the key to transformation. That does not necessarily mean more information, better statistics, bigger databases, or the World Wide Web, though all of these may play a part. It means relevant, compelling, select, powerful, timely, accurate information flowing in new ways to new recipients, carrying new content, suggesting new rules and goals (rules and goals which are themselves information). Any system will behave differently when its information flows are changed. The policy of Glasnost, for example, the simple opening of information channels that had long been closed in the Soviet Union, guaranteed the rapid transformation of Eastern Europe beyond anyone's expectation. The old system had been held in place by tight control of information. Letting go of that control triggered total system restructuring (turbulent and unpredictable, but inevitable).

Second, systems strongly resist changes in their information flows, especially in their rules and goals. It is not surprising that those who benefit from the current system actively oppose such revision. Entrenched political, economic, and religious cliques can constrain almost entirely the attempts of an individual or small group to operate by different rules or to attain goals different from those sanctioned by the system. Innovators can be ignored, marginalized, ridiculed, denied promotions or resources or public voices. They can be literally or figuratively snuffed out.

Only innovators, however—by perceiving the need for new information, rules, and goals, communicating about them, and trying them out—can make the changes that transform systems. This important point is expressed clearly in a quote that is widely attributed to Margaret Mead, "Never doubt that a small group of thoughtful, committed citizens can change the world. Indeed, it's the only thing that ever has."

We have learned the hard way that it is difficult to live a life of material moderation within a system that expects, exhorts, and rewards consumption. But one can move a long way in the direction of moderation. It is not easy to use energy efficiently in an economy that produces energy-inefficient products. But one can search out, or if necessary invent, more efficient ways of doing things, and in the process make those ways more accessible to others.

Above all, it is difficult to put forth new information in a system that is structured to hear only old information. Just try, sometime, to question in public the value of more growth, or even to make a distinction between growth and development, and you will see what we mean. It takes courage and clarity to challenge an established system. But it can be done.

In our own search for ways to encourage the peaceful restructuring of a system that naturally resists its own transformation, we have tried many tools. The obvious ones are . . . rational analysis, data, systems thinking, computer modeling, and the clearest words we can find. Those are tools that anyone trained in science and economics would automatically grasp. Like recycling, they are useful and necessary, and they are not enough.

We don't know what will be enough. But . . . here are . . . five other tools we have found helpful. We introduced and discussed this list for the first time in our 1992 book. Our experience since then has affirmed that these five tools are not optional; they are essential characteristics for any society that hopes to survive over the long term. We present them here . . . "not as the ways to work toward sustainability, but as some ways" (Meadows et al. 1992).

"We are a bit hesitant to discuss them," we said in 1992, "because we are not experts in their use and because they require the use of words that do not come easily from the mouths or word processors of scientists. They are considered too 'unscientific' to be taken seriously in the cynical public arena." What are the tools we approached so cautiously?

They are: visioning, networking, truth-telling, learning, and loving. It seems like a feeble list, given the enormity of the changes required. But each of these exists within a web of positive loops. Thus their persistent and consistent application initially by a relatively small group of people would have the potential to produce enormous change—even to challenge the present system, perhaps helping to produce a revolution.

"The transition to a sustainable society might be helped," we said in 1992, "by the simple use of words like these more often, with sincerity and without apology, in the information streams of the world." But we used them with apology ourselves, knowing how most people would receive them.

Many of us feel uneasy about relying on such "soft" tools when the future of our civilization is at stake, particularly since we do not know how to summon them up, in ourselves or in others. So we dismiss them and turn the conversation to recycling or emission trading or wildlife preserves or some other necessary but insufficient part of the sustainability revolution—but at least a part we know how to handle.

So let's talk about the tools we don't yet know how to use, because humanity must quickly master them.

Visioning

Visioning means imagining, at first generally and then with increasing specificity, what you really want. That is, what you really want, not what someone has taught you to want, and not what you have learned to be willing to settle for. Visioning means taking off the constraints of "feasibility," of disbelief and past disappointments, and letting your mind dwell upon its most noble, uplifting, treasured dreams.

Some people, especially young people, engage in visioning with enthusiasm and ease. Some find the exercise of visioning frightening or painful, because a glowing picture of what could be makes what is all the more intolerable. Some people never admit their visions, for fear of being thought impractical or "unrealistic." They would find this paragraph uncomfortable to read, if they were willing to read it at all. And some people have been so crushed by their experience that they can only explain why any vision is impossible. That's fine; skeptics are needed too. Vision needs to be disciplined by skepticism. We should say immediately, for the sake of the skeptics, that we do not believe vision makes anything happen. Vision without action is useless. But action without vision is directionless and feeble. Vision is absolutely necessary to guide and motivate. More than that, vision, when it is widely shared and firmly kept in sight, does bring into being new systems.

We mean that literally. Within the limits of space, time, materials, and energy, visionary human intentions can bring forth not only new information, new feedback loops, new behavior, new knowledge, and new technology, but also new institutions, new physical structures, and new powers within human beings. Ralph Waldo Emerson recognized this profound truth 150 years ago:

> Every nation and every man instantly surround themselves with a material apparatus which exactly corresponds to their moral state, or their state of thought. Observe how every truth and every error, each a thought of some man's mind, clothes itself with societies, houses, cities, language, ceremonies, newspapers. Observe the ideas of the present day . . . see how each of these abstractions has embodied itself in an imposing apparatus in the community, and how timber, brick, lime, and stone have flown into convenient shape, obedient to the master idea reigning in the minds of many persons. . . .
>
> It follows, of course, that the least change in the man will change his circumstances; the least enlargement of ideas, the least mitigation of his feelings in respect to other men . . . would cause the most striking changes of external things. (Emerson, 1838)

A sustainable world can never be fully realized until it is widely envisioned. The vision must be built up by many people before it is complete and compelling. As a way of encouraging others to join in the process, we'll list here some of what we see when we let ourselves imagine a sustainable society we would like to live in—as opposed to one we would be willing to settle for. This is by no means a definitive list. We include it here only to invite you to develop and enlarge it.

- Sustainability, efficiency, sufficiency, equity, beauty, and community as the highest social values.
- Material sufficiency and security for all. Therefore, by individual choice as well as communal norms, low birth rates and stable populations.
- Work that dignifies people instead of demeaning them. Some way of providing incentives for people to give their best to society and to be rewarded for doing so, while ensuring that everyone will be provided for sufficiently under any circumstances.
- Leaders who are honest, respectful, intelligent, humble, and more interested in doing their jobs than in keeping their jobs, more interested in serving society than in winning elections.
- An economy that is a means, not an end; one that serves the welfare of the environment, rather than vice versa.
- Efficient, renewable energy systems.
- Efficient, closed-loop materials systems.

• Technical design that reduces emissions and waste to a minimum, and social agreement not to produce emissions or waste that technology and nature can't handle.

• Regenerative agriculture that builds soils, uses natural mechanisms to restore nutrients and control pests, and produces abundant, uncontaminated food.

• The preservation of ecosystems in their variety and human cultures living in harmony with those ecosystems; therefore, high diversity of both nature and culture, and human appreciation for that diversity.

• Flexibility, innovation (social as well as technical), and intellectual challenge. A flourishing of science, a continuous enlargement of human knowledge.

• Greater understanding of whole systems as an essential part of each person's education.

• Decentralization of economic power, political influence, and scientific expertise.

• Political structures that permit a balance between short-term and long-term considerations, some way of exerting political pressure now on behalf of our grandchildren.

• High skills on the part of citizens and governments in the arts of nonviolent conflict resolution.

• Media that reflect the world's diversity and at the same time unite cultures with relevant, accurate, timely, unbiased, and intelligent information, presented in its historic and whole-system context.

• Reasons for living and for thinking well of oneself that do not involve the accumulation of material things.

Networking

We could not do our work without networks. Most of the networks we belong to are informal. They have small budgets, if any, and few of them appear on rosters of world organizations.[1] They are almost invisible, but their effects are not negligible. Informal networks carry information in the same way as formal institutions do, and often more effectively. They are the natural home of new information, and out of them new system structures can evolve.[2]

Some of our networks are very local, some are international. Some are electronic, some involve people looking each other in the face every day. Whatever their form, they are made up of people who share a common interest in some aspect of life, who stay in touch and pass around data and tools and ideas and encouragement, who like and respect and support each other. One of the most important purposes of a network is simply to remind its members that they are not alone.

A network is non-hierarchical. It is a web of connections among equals, held together not by force, obligation, material incentive, or social contract, but by shared values and the understanding that some tasks can be accomplished together that could never be accomplished separately.

We know of networks of farmers who share organic pest control methods. There are networks of environmental journalists, "green" architects, computer modelers, game designers, land trusts, consumer cooperatives. There are thousands and thousands of networks that developed as people with common purposes found each other. Some networks become so busy and essential that they evolve into formal organizations with offices and budgets, but most come and go as needed. The advent of the World Wide Web certainly has facilitated and accelerated the formation and maintenance of networks.

Networks dedicated to sustainability at both the local and the global levels are especially needed to create a sustainable society that harmonizes with local ecosystems while keeping itself within global limits. About local networks we can say little here; our localities are different from yours. One role of local networks is to help reestablish the sense of community and relation to place that has been largely lost since the Industrial Revolution. . . .

Truth-Telling

We are no more certain of the truth than anyone is. But we often know an untruth when we hear one. Many untruths are deliberate, understood as such by both speaker and listeners. They are put forth to manipulate, lull, or entice, to postpone action, to justify self-serving action, to gain or preserve power, or to deny an uncomfortable reality.

Lies distort the information stream. A system cannot function well if its information streams are corrupted by lies. One of the most important tenets of systems theory . . . is that information should not be distorted, delayed, or sequestered.

"All of humanity is in peril," said Buckminster Fuller, "if each one of us does not dare, now and henceforth, always to tell only the truth and all the truth, and to do so promptly—right now" (Buckminster Fuller 1981). Whenever you speak to anyone, on the street, at work, to a crowd, and especially to a child, you can endeavor to counter a lie or affirm a truth. You can deny the idea that having more things makes one a better person. You can question the notion that more for the rich will help the poor. The more you can counter misinformation, the more manageable our society will become.

Here are some common biases and simplifications, verbal traps and popular untruths that we run into frequently in discussing limits to growth. We think they need to be pointed out and avoided, if there is ever to be clear thinking about the human economy and its relationship to a finite earth.

Not: A warning about the future is a prediction of doom.
But: A warning about the future is a recommendation to follow a different path.

Not: The environment is a luxury or a competing demand or a commodity that people will buy when they can afford it.
But: The environment is the source of all life and every economy. Opinion polls typically show that the public is willing to pay more for a healthy environment.

Not: Change is sacrifice, it should be avoided.
But: Change is challenge, and it is necessary.

Not: Stopping growth will lock the poor in their poverty.
But: It is the avarice and indifference of the rich, which locks the poor into poverty. The poor need new attitudes among the rich; then there will be growth specifically geared to serve their needs.

Not: Everyone should be brought up to the material level of the richest countries.
But: There is no possibility of raising material consumption levels for everyone to the levels now enjoyed by the rich. Everyone should have his or her fundamental material needs satisfied. Material needs beyond this level should be satisfied only if it is possible, for all, within a sustainable ecological footprint.

Not: All growth is good, without question, discrimination, or investigation.
Nor: All growth is bad.

But: What is needed is not growth, but development. Insofar as development requires physical expansion, it should be equitable, affordable, and sustainable, with all real costs counted.

Not: Technology will solve all problems.
Nor: Technology does nothing but cause problems.
But: We need to encourage technologies that will reduce the ecological footprint, increase efficiency, enhance resources, improve signals, and end material deprivation.
And: We must approach our problems as human beings and bring more to bear on them than just technology.

Not: The market system will automatically bring us the future we want.
But: We must decide for ourselves what future we want. Then we can use the market system, along with many other organizational devices, to achieve it.

Not: Industry is the cause of all problems, or the cure.
Nor: Government is the cause or the cure.
Nor: Environmentalists are the cause or the cure.
Nor: Any other group (economists come to mind) is the cause or the cure.
But: All people and institutions play their role within the large system structure. In a system that is structured for overshoot, all players deliberately or inadvertently contribute to that overshoot. In a system that is structured for sustainability, industries, governments, environmentalists, and most especially economists will play essential roles in contributing to sustainability.

Not: Unrelieved pessimism.
Nor: Sappy optimism.
But: The resolve to tell the truth about both the successes and failures of the present and the potentials and obstacles in the future.
And above all: The courage to admit and bear the pain of the present, while keeping a steady eye on a vision of a better future.

Not: The World model, or any other model, is right or wrong.
But: All models, including the ones in our heads, are a little right, much too simple, and mostly wrong. How do we proceed in such a way as to test our models and learn where they are right and wrong? How do we

speak to each other as fellow modelers with an appropriate mixture of skepticism and respect? How do we stop playing right/wrong games with each other and start designing right/wrong tests for our models against the real world?

That last challenge, sorting out and testing models, brings us to the topic of learning.

Learning

Visioning, networking, and truth-telling are useless if they do not inform action. There are many things to *do* to bring about a sustainable world. New farming methods have to be worked out. New businesses have to be started and old ones have to be redesigned to reduce their footprint. Land has to be restored, parks protected, energy systems transformed, international agreements reached. Laws have to be passed and others repealed. Children have to be taught and so do adults. Films have to be made, music played, books published, websites established, people counseled, groups led, subsidies removed, sustainability indicators developed, and prices corrected to portray full costs.

All people will find their own best role in all this doing. We wouldn't presume to prescribe a specific role for anyone but ourselves. But we would make one suggestion: Whatever you do, do it humbly. Do it not as immutable policy, but as experiment. Use your action, whatever it is, to learn.

The depths of human ignorance are much more profound than most of us are willing to admit. This is especially at a time when the global economy is coming together as a more integrated whole than it has ever been, when that economy is pressing against the limits of a wondrously complex planet, and when wholly new ways of thinking are called for. At this time, no one knows enough. No leaders, no matter how authoritative they pretend to be, understand the situation. No policy should be imposed wholesale upon the whole world. If you cannot afford to lose, do not gamble.

Learning means the willingness to go slowly, to try things out, and to collect information about the effects of actions, including the crucial but not always welcome information that the action is not working. One can't learn without making mistakes, telling the truth about them, and moving on. Learning means exploring a new path with vigor and courage, being open to other peoples' explorations of other paths, and being willing to switch paths if one is found that leads more directly to the goal.

The world's leaders have lost both the habit of learning and the freedom to learn. Somehow a political system has evolved in which the voters expect leaders to have all the answers, that assigns only a few people to be leaders, and that brings them down quickly if they suggest unpleasant remedies. This perverse system undermines the leadership capacity of the people and the learning capacity of the leaders.

It's time for us to do some truth-telling on this issue. The world's leaders do not know any better than anyone else how to bring about a sustainable society; most of them don't even know it's necessary to do so. A sustainability revolution requires each person to act as a learning leader at some level, from family to community to nation to world. And it requires each of us to support leaders by allowing them to admit uncertainty, conduct honest experiments, and acknowledge mistakes.

No one can be free to learn without patience and forgiveness. But in a condition of overshoot, there is not much time for patience and forgiveness. Finding the right balance between the apparent opposites of urgency and patience, accountability and forgiveness is a task that requires compassion, humility, clear-headedness, honesty and—that hardest of words, that seemingly scarcest of all resources—love.

Loving

One is not allowed in the industrial culture to speak about love, except in the most romantic and trivial sense of the word. Anyone who calls upon the capacity of people to practice brotherly and sisterly love, love of humanity as a whole, love of nature and of our nurturing planet, is more likely to be ridiculed than to be taken seriously. The deepest difference between optimists and pessimists is their position in the debate about whether human beings are able to operate collectively from a basis of love. In a society that systematically develops individualism, competitiveness, and short-term focus, the pessimists are in the vast majority. Individualism and short-sightedness are the greatest problems of the current social system, we think, and the deepest cause of unsustainability. Love and compassion institutionalized in collective solutions is the better alternative. A culture that does not believe in, discuss, and develop these better human qualities suffers from a tragic limitation in its options. "How good a society does human nature permit?" asked psychologist Abraham Maslow. "How good a human nature does society permit?" (Maslow 1971).

The sustainability revolution will have to be, above all, a collective transformation that permits the best of human nature, rather than the worst, to be expressed and nurtured. Many people have recognized that necessity and that opportunity. For example, John Maynard Keynes wrote in 1932:

> The problem of want and poverty and the economic struggle between classes and nations is nothing but a frightful muddle, a transitory and unnecessary muddle. For the Western World already has the resource and the technique, if we could create the organization to use them, capable of reducing the Economic Problem, which now absorbs our moral and material energy, to a position of secondary importance. . . . [Thus the] day is not far off when the Economic Problem will take the back seat where it belongs, and . . . the arena of the heart and head will be occupied . . . by our real problems—the problems of life and of human relations, of creation and behaviour and religion (Keynes, 1932). . . .

Is anything we have advocated . . . , from more resource efficiency to more compassion, really possible? Can the world actually ease down below the limits and avoid collapse? Can the human footprint be reduced in time? Is there enough vision, technology, freedom, community, responsibility, foresight, money, discipline, and love, on a global scale?

Of all the hypothetical questions we have posed in this book, these are the most unanswerable, though many people will pretend to answer them. Even we—your authors—differ among ourselves when tallying the odds for and against. The ritual cheerfulness of many uninformed people, especially world leaders, would say the questions are not even relevant; there are no meaningful limits. Many of the informed are infected with the deep cynicism that lies just under the ritual public cheerfulness. They would say that there are severe problems already, with worse ones ahead, and that there's not a chance of solving them.

Both of those answers are based, of course, on mental models. The truth of the matter is that *no one knows*.

We have said many times . . . that the world faces not a preordained future, but a choice. The choice is between different mental models, which lead logically to different scenarios. One mental model says that this world for all practical purposes has no limits. Choosing that mental model will encourage extractive business as usual and take the human economy even further beyond the limits. The result will be collapse.

Another mental model says that the limits are real and close, and that there is not enough time, and that people cannot be moderate or responsible or compassionate. At least not in time. That model is self-fulfilling. If the world's people choose to believe it, they will be proven right. The result will be collapse.

A third mental model says that the limits are real and close and in some cases below our current levels of throughput. But there is just enough time, with no time to waste. There is just enough energy, enough material, enough money, enough environmental resilience, and enough human virtue to bring about a planned reduction in the ecological footprint of humankind: a sustainability revolution to a much better world for the vast majority.

That third scenario might very well be wrong. But the evidence we have seen, from world data to global computer models, suggests that it could conceivably be made right. There is no way of knowing for sure, other than to try it.

Notes

1. Examples of networks known to the authors and in their field of interest are the Balaton Group (www.unh.edu/ipssr/Balaton.html), Northeast Organic Farming Association (NOFA), Center for a New American Dream (CNAD; www.newdream.org), Greenlist (www.peacestore.us/Public/Greenlist), Greenclips (www.greenclips.com), Northern Forest Alliance (www.northernforestalliance.org), Land Trust Alliance (www.lta.org), International Simulation and Gaming Association (ISAGA; www.isaga.info), and Leadership for Environment and Development (LEAD).

2. Such an intermediate step is illustrated by ICLEI, an international association of (currently 450) local governments implementing sustainable development. See www.iclei.org.

References

Buckminster Fuller, R. 1981. *Critical Path*. New York: St. Martin's Press.

Emerson, Ralph Waldo. 1838. Lecture on "War," delivered in Boston, March 1838. Reprinted in *Emerson's Complete Works*, vol. 11. Boston: Houghton Mifflin, 177.

Keynes, J. M. 1932. Foreword to *Essays in Persuasion*. New York: Harcourt Brace.

Maslow, Abraham. 1971. *The Farthest Reaches of Human Nature*. New York: Viking Press.

Meadows, Donella, Dennis Meadows, and Jørgen Randers. 1992. *Beyond the Limits*. Post Mills, VT: Chelsea Green Publishing Company.

Worster, Donald, editor. 1988. *The Ends of the Earth*. Cambridge: Cambridge University Press, 11–12.

25

Downshift/Upshift: Our Choice*

Raymond De Young and Thomas Princen

In this book we have attempted to facilitate a conversation based on the premise that irreversible environmental disruption is imminent and that net energy availability is in decline. From this premise we have argued that society will be undergoing unprecedented changes. People will have to be creative, but creativity won't be enough. A change in worldview at both the individual cognitive level and the societal political level will also be needed. So there will be a downshift, a monumental change from the historical expectation of ever-increasing material and energy abundance to a future expectation of a steady state at lower levels of consumption. Human behavior must adjust; how much no one knows. The following observations lead us to expect it will be a lot.

1. The world has a century-plus habit of using cheap fossil fuels (some 80 percent of all energy used worldwide is derived from fossil fuels, and most of that is consumed in the North).
2. Humans have consumed resources and dumped their wastes almost costlessly (most resource wastes disperse unseen into the atmosphere).
3. Most of the world's infrastructure is heavily dependent on cheap and easy fossil fuels (e.g., pipelines, power lines, highways, buildings).
4. World economies, whether capitalist or socialist, developed or developing, are designed for and require continuous growth.
5. The world has finite resources.

With these observations in mind, we fully expect that Northern, high-consuming societies will, one way or another, transition to dramatically reduced rates of consumption, at least until they find operating ranges where consumption of natural resources matches regeneration.

For the immediate future, the magnitude of the needed change is less important than its direction. For the past 150 years, consumption has shifted up—always more and bigger and faster. During this relatively brief time in human history, perhaps an era to be named by future historians as the efflorescence of oil, society has experienced increasing material use as a result of its core economic idea that everything must grow to remain healthy. The obvious result has been ever more machines, more goods and services, more transport, more discretionary spending. On top of that, cultural ideas of progress and development have contained a similar growth imperative, a sense of individual entitlement to have more.

Now the upshift can no longer be taken for granted. Soon the direction—specifically the material direction—will be downward: less energy, less water, less food, less timber, less fish. And less not because people have made the conscious choice to reduce consumption, but because the emerging biophysical reality makes current levels of consumption impossible. The era of cheap fossil fuels—cheap in finding, extracting, refining, transporting, consuming, and disposing of their wastes—is coming to an end.

While the direction of material consumption will be down, in other areas of life the direction may be up—more time, more choice, more personal well-being, more social cohesion, and more overall security. This new expectation, the "upside of down,"[1] follows from the plaguing observation that the current order, for all its goods and services, actually has some nontrivial bads and disservices—the regularity (even tolerance) of transportation deaths, insecure families, stressed communities, and, perhaps the mother of all disservices, imminent climate destabilization. The flip side of the material downshift, then, is nonmaterial upshift, an expectation of more spare time, more conviviality, more connection to nature, and better food. Indeed, the upshift may generate high levels of well-being because it has the potential to make people feel truly needed by their communities and truly secure in their lives.

This view, while consistent with the contributions to this book, is hardly mainstream thinking and may not be for some time.[2] It simply may be too difficult for some people to accept. That said, as researchers and teachers, our sense is that a growing segment of younger generations find the idea of a downshift in consumption and a corresponding upshift in quality of life rather easy to digest. They are witness to decades of ever more frantic efforts to reduce resource consumption, to discover technological fixes to environmental calamities, to develop substitutes for fossil fuels, and to preserve forests. Perhaps for this reason, they have little difficulty understanding the need for a shift to a new normal, one more frugal and spare, one not unlike that of older generations.

But localization is about more than just understanding the need for dramatic change; it is also about starting the transition while there are still options. Certainly, the downshift will happen whether we are prepared or not because, as Herman Daly reminds us, "sustainable growth is impossible."[3] But if we prepare now for living with less at a time when there is still enough, then we—individuals and society as a whole—may experience a peaceful, democratic, just, and resilient transition, thus unleashing the upside of the downshift.

The Downshift Moment

In studying and teaching localization, we have observed that after study or reflection or both, people often come to a "downshift moment." This can happen at a point in knowledge accumulation when inert facts, known but not owned, become part of one's reality. Evidence, trends, and causal relations coalesce into a new mental map. In the case of transition and downshift, the initial facts are largely about the material—natural resources and wastes on the one hand, money, credit, debt, and insurance on the other. At other times a downshift moment happens when people "face the facts" about *change*, about bewildering trends, declining availability, displaced costs that finally come due, discontinuities in supply and in the environment. But, we surmise, the downshift moment is more than a matter of knowledge accumulation; it is also a matter of clarity emerging from the realization that in one's lifetime, steady material decline will become the new order. The downshift moment can even be a sensuous moment underlain by strong but mixed emotions. One can almost feel the difference between the experiences and expectations of the cheap-energy era and those of the coming transition and the steady state beyond. Recalling with pleasure an electrifying lifestyle while grieving its loss will certainly be part of the experience. Uncertainty and anxiety will surely accompany such a sweeping transition, but anticipation and fascination can also be expected, feelings borne by the possibility of meaningful challenge and opportunity.

Author and organizer Jerry Mander approaches the same issue from a slightly different angle: "Utopian societies," he says, are places where "Everything is planned. Everything figured out. Everything created."[4] For the human mind—which is adapted to exploration and motivated to increase competence—such a place would be dismal. Therefore, localization should not be approached as a utopia-seeking process. Rather, it should be approached as an opportunity to draw people's finest behavior, not by making life easier, but by acknowledging the hard work ahead and participating fully in the crafting of appropriate responses. Localization is a new struggle, perhaps a battle, but one that may well be led by ordinary people not experts, by people who, although sometimes scared, will do the work needed, one bit at a time (chapter 16). Aiding their efforts are some well-established facts about what people need to do to thrive

1. Pursue meaningfulness, not pleasure.
2. Seek clarity and clear-headedness, not convenience.
3. Aim for exploration and problem solving (what our minds are adapted for), not affluence (what our minds and our institutions are ill adapted for).
4. Build competence through engaged, self-directed occupation, not through leisure activities.

As observers of social action across the spectrum, we find that these behaviors are ubiquitous wherever people thrive. Because they pertain equally to rich and poor, powerful and weak, we expect they will be prominent in positive localization.

What all this says is that localization does not accept the dominant assumption that increasing affluence is an unmitigated good, nor that lowering material wealth necessarily decreases well-being. In fact, affluence and the well-being presumed to go with it can be the product of brutal working hours and workplace competition, inhumane levels of stress and uncertainty, and destructive interactions with natural systems. Even with the best of intentions, the pursuit of individual wealth can and often does do harm to people, institutions, higher values, and natural systems. In contrast, a life pattern that intentionally pursues less material wealth may allow for more time with family and friends, more opportunity to volunteer in one's community, more occasion to revitalize body, mind, and spirit, and more fulfilling relations with the natural world.

Ultimately, it would be nice if we had a theory of localization to offer communities in transition, something on the order of natural selection in biology or the price mechanism in economics. But we do not. Though much localizing is already happening, ours is a new envisioning exercise. It includes ongoing attempts to cope with inescapable trends in topsoil, freshwater, and fossil fuel depletion. And it includes a normative question: What should society do to steer its adaptation away from negative localization (a transition dominated by warlords and survivalists) and toward positive localization (a transition that is peaceful, democratic, just, and resilient)?

Although we cannot offer a theory, we can offer a few principles, as preliminary and incomplete as they are at this stage. The reader will see that these principles are not only different from the prevailing principles of social organization—namely growth, efficiency, convenience, and consumer sovereignty—but they are largely antithetical to such principles. They are grounded in the functioning of communities—social (human) and ecological (nonhuman biological)—and their interactions. If there is an order to these principles, it is from inward looking (engaging in one's own locality) to outward looking (engaging the surrounding institutional and cultural environment in which the locality must function). That said, we prefer to view these as elements of localization that function simultaneously, some overlapping, most reinforcing.

There will inevitably be variation in how communities would use these principles. Responses that are durable at greatly reduced levels of energy and resource use will be selected for. We contend that positive localization can be among those selected responses and that their participants will thrive and their communities will prosper. We offer these principles as tools for crafting successful responses.

Many of the principles below emerged from discussions between us editors and among our students over several years of mutual exploration. Others come directly from the contributions to this book. None are perfect but we hope that all are suggestive and provocative. We also hope that readers will come up with their own principles, share them with activists and scholars, and build a body of tentative propositions that someday someone might call a theory of localization.

Some Principles for Localization

1. *Diversity-of-localities principle* Each locality should solve as many of its own problems as possible and do so in ways suited to its own biophysical and social conditions. Self-directed, uniquely determined problem solving is likely to be a source of creativity, not just in provisioning basics (e.g., food, water, shelter, transportation), but in fashioning new institutions and new cultural practices, including those of participatory democracy, ecological sustainability, diversity, and equity.

Historian Rudolf Rocker argues that the creativity and vitality of ancient Greece stemmed in part from its multiplicity of uniquely organized localities: "It was this healthy decentralization, this internal separation of Greece into hundreds of little communities, tolerating no uniformity, which constantly aroused the mind."[5] This suggests that a process of localizing, far from being a force of repression, retrenchment, or going backward, has been, and can be, a force for creativity and reinvention. And that creativity can be directed as much to designing ecologically resilient and equitable institutions as to democracy and artistic expression. What's more, diversified localities match diversified ecosystems. Because no two assemblages of land, water, and biota are alike, an ecologically adapted society would be one whose localities are locally adapted, each uniquely situated in place. The social requisites for this biophysical adaptation would be decentralized and place-based decision making (see below), especially for the basics: food, water, shelter, transportation.

Corollary—Oxygen-mask principle People in transition should organize their own localities before reorganizing their nation or the world. Only when there are surpluses locally—in water, food, even time and skills—does one look outward. If higher levels of authority are needed to ensure local provisioning, then one organizes at those levels. Otherwise, one looks first inward, drawing on local capacities to provide for local infrastructure and needs.

In the era of cheap and easy oil, it may have made sense for local economies to be tightly interdependent with national and global economies. Without cheap oil, it makes more sense, from a resilience perspective, to secure the basics. The sense making of the petroleum era was narrowly economic (if GDP or trade increased, it was good); the sense making of the postpetroleum era is the predictability and security of basic needs (it is good to know clean water and healthy food are reliably available).

2. *Semipermeable-boundaries principle* Localizers should deliberately construct boundaries for the two-way flow of materials, money, people, and ideas. That flow must be managed so as to maintain the integrity of the local system. Where flows cannot be so managed, higher levels of organization must be established (see the oxygen-mask principle above and the subsidiarity principle below).

David Korten writes in his "Joys of Earth Community" that "life's processes of self-renewal depend on a managed exchange with its environment. Therefore, each organism's boundary membrane must be permeable, and what flows through that membrane in both directions must be subject to management by the organism." In a similar vein, one of us (Princen) has written in the context of direct resource use that "much as the body's cells protect their interior organelles at the same time that they exchange gases and nutrients with their environment, boundaries are never rigid or impassable. And just as humans and other animals benefit genetically from migration and crossbreeding, resource communities benefit from the movements of seed and prey and new participants. Boundaries are necessary but they must be permeable. What is more . . . they must be selectively permeable."[6]

3. *Place-based decision-making principle* When critical life-support systems are at risk, key decisions should reside with those who demonstrate a connection and commitment to place, not with those who are placeless.

This is a residential principle in that people who live and work in a community are more likely to have local knowledge and represent community values. It is also a positive dependency principle in that such people are dependent on the coherence and durability of their own community. They have a major stake in the community, are likely to resist top-down decision making, and thus are more likely to factor in conditions, opportunities, and constraints best known to community members. For instance, local farmers know, through crop yields, when they are pushing the soil too hard. They may not know the ultimate or even proximate cause, but they know their soil's limits, and thus their knowledge constitutes a constraint on local food and fiber production. It is precisely such built-in dependencies and constraints that placeless decision making in the national and global economies lacks.

Put differently, place-based problem solving should supersede private and distant decision making (e.g., by a federal government or transnational corporation). It should engage as many of the affected members of the community as possible.

This proposition follows from the psychology of motivation and, especially, competence, as it relates to community and place. Mary Midgley contrasts the reductionist view of motivation in "economic man" (i.e., "utility," that all-purpose measure of satisfaction that, in practice, is immeasurable) with that of those who actually live in, construct, and adapt to place: "The attempt to reduce plurality of motivation to abstract unity ['utiles' or money] shrinks the essential self to a wizened old nut, a bare intellectual center of choice, unattached to particular people and things and equally capable—if its one abstract need is met—of living anywhere."[7] It is an extension of her idea that the consumer and the investor in today's economy have no connection to place; they just buy and sell.

Regarding competence, psychologist Stephen Kaplan sees helplessness as a pivotal issue in environmentally responsible behavior.[8] Some social structures tend to foster such helplessness: monarchies, dictatorships, and consumer economies, for instance. As for consumer economies, people, as consumers, can do little more than buy better; they often can only choose from among that which is offered to them. By contrast, the more localized the economy the more the integrity of families, communities, and natural systems is the central economic concern and the more local competence is called for and rewarded.

Corollary 1—Internal dependence In encouraging economic development, government should give preference to businesses that can demonstrate a structural dependence on the local economy. When the local economy does well such businesses do well, and when it does poorly local businesses do poorly.

This principle enacts a counterweight to mobile capital, to the tendency among placeless corporations, investors, and manufacturers to pull up stakes when a community declines rather than investing in the stability of that community. The requisite structural dependence could be physical—farming a designated stretch of land for a well-suited crop, for example—or it could be property and residence based—owners live and work locally as a condition of government support.

To implement this principle, some measure of the local economy is needed, something that is not simply a fraction of the country's GDP, let alone of gross revenues or market share value. With an internal dependence measure, a business would have to show that if the local economy thrives, that business does likewise, and, maybe most importantly, that its fortunes rise and fall with the local economy. Community Supported Agriculture organizations (CSAs) might provide a model (see chapter 9), as might local credit unions, as well as locally owned, independent banks and retailers (see chapter 7).[9]

Corollary 2—Minimal money Participants in a localizing process should minimize the role of abstract financial instruments, perhaps even money itself, especially to the extent such instruments connect to large-scale, centralized systems (e.g., the global banking system) and disconnect people from the communities and ecosystems they depend on.

4. *Subsidiarity principle* Decision making should be at the lowest possible level—that is, as close to affected peoples and their resources as possible. Water supply and sanitation, for instance, should be arranged by those who enjoy the services and pay for them; most often this will occur at small scales, within a watershed, and bounded politically and financially. By contrast, reduction in greenhouse gases should be arranged globally, by the international community of nations. Both examples conform to the subsidiarity principle[10] and both have ecological content.

5. *Subsistence principle* A locality should subsist, as much as possible, from the production of that locality and its immediate region, and should take measures to secure such production for the long term. The objective is human security, material and social, rather than revenue generation or profit.

In food systems, the well-known fragility of monocultures, concentrated ownership, and distanced supply networks must give way to diverse crops as well as diverse methods and modes of production. In general, subsistence production of food is likely to

i. Be best adapted to local conditions and needs.
ii. Minimize transport and other energy costs (and thus better prepare the community for diminishing fossil fuels and available net energy).
iii. Allow export when there are surpluses and aid when there are emergencies.
iv. Enhance the long-term food security of the locality.

In short, subsistence-oriented production is inherently diversified, which increases system resilience. Moreover, local finance is likely to be more resilient because money is more liable to stay in the community longer than with large-scale food systems. There are some who will protest that a subsistence principle is protectionist. Wendell Berry says "that is exactly what it is. It is a protectionism that is just and sound, because it protects local producers and is the best assurance of adequate supplies to local consumers." He then makes a crucial distinction. A just protectionism is "the best guarantee of giveable or marketable surpluses. This kind of protection is not 'isolationism.' "[11]

6. *Artisanal principle* When productive, locally oriented enterprises—businesses, private or public or common, profit-making or not—develop products for exchange, that production should

i. Find a scale of production well suited to a locality or collection of localities—that is, it should be at a scale that is minimally dependent on external supplies and markets and maximally dependent on local labor, natural resources, and finances.

ii. Build in brakes on expansion or contraction that otherwise militates against appropriate scale.

iii. Develop and teach skills first, technologies second, and minimize technologies that replace people and their special skills.

iv. Promote "inefficiencies," measures that intentionally limit short-term gain in exchange for long-term secure production and, hence, human security.

v. Find seasonalities, intermittencies, and constructive ambiguities to build variability and innovation into the local economy.

vi. Minimize commodification and the role of money, instead increasing relations—human, natural, communal—that are simultaneously social and commercial, problem solving and problem defining.

7. *Land-access principle* Access to land for self-reliant provisioning (e.g., food, water, wastewater treatment, fuel) is a fundamental right and should be institutionalized at all levels of government.

This principle suggests that land for self-provisioning is a public trust and should be so treated, legally, politically, and economically;[12] it should not enter into market exchange. It also implies that where such access currently exists, no actions should be taken to diminish those rights and no infrastructure should be built (e.g., dams, highways), no policy adopted, no taxes levied or subsidies offered that cut off such land access.

One implication of this principle is that communities, in anticipation of increasing demand for local food, should identify productive land and clean water sources whether public, private, or common. Then they should begin to acquire public trust rights for citizen access, with appropriate rules and procedures.

8. *Clay-road principle* Localizers should build resilient, locally adaptable systems of transport, water, food, and energy. The superstructure (e.g., cars, trucks, buses, railroads, boats, barges) is less critical than the substrate (e.g., roadbed, waterway). Substrates that are readily built and maintained with local materials and skills should be given preference. The likely results are slower transport, less exotic foods, and less mobility overall, on the one hand, and more reliable systems on the other.

Taking transportation as an example, the current highway system is fragile. It will deteriorate quickly if its maintenance schedule is not followed. Since the highway system is highly centralized, in constant need of maintenance and fossil fuel dependent (even aside from the vehicles' gasoline consumption), it is thus a brittle system. A localized transport system, by contrast, would be the opposite: adapted to local conditions, dependent primarily on local materials and energy sources. Since such systems are unlikely to mimic the national highway system, and may involve conflicting standards, they are likely to result in slower transport. The challenge for localizers is to make them resilient—that is, physically and socially robust.

A development project in Malawi offers a case in point. A major hindrance to development in rural villages is access to markets. Because few people have cars, modern highways are irrelevant. Through trial and error, development specialists and villagers discovered that an appropriate road could be made by villagers from local materials. A porous bed of gravel was overlain with tamped down clay. Use of the road further tamped down and smoothed the clay. Bicycles, locally produced and ubiquitous, and carts traverse this surface readily. When the surface cracks, rains seep into the cracks, and the clay absorbs the water and swells up, sealing the crack.[13] This particular feature—self-correcting, usable even when damaged—suggests a corollary principle, the escalator principle: in a building of modest height, escalators are more appropriate than elevators because, when the electricity goes off, escalators become stairs.

9. *Adaptive-muddling principle* To increase the probability of success, localizers should deliberately experiment. Proponents should constantly tinker with new institutional forms, metaphors, norms, behaviors, and principles. Although current analytical tools help make sense of the past (e.g., how we got to this state of fossil fuel dependence) and the present (e.g., the status of the environmental predicament) and can extrapolate recent trends into the future, they cannot determine which paths into the future will be meaningful and effective. For this localizers should adopt an adaptive, experimental approach. Problem solving would aim for a plurality of solutions, not the one right solution. Emerging plans, policies, and procedures would be viewed as hypotheses in constant need of testing. Or, as author and community organizer Pat Murphy puts it, we need to "make a lot of mistakes quickly."[14]

An experimental approach to social change under conditions of great uncertainty and grave stakes (a premise of this book) might start with small steps. As anthropologist and political scientist James Scott advises with respect to interventions for economic development, "Prefer wherever possible to take a small step, stand back, observe, and then plan the next small move."[15] Still, with the challenges outlined in this book, a sequence of small steps may not be sufficient. We may need many steps, in many directions, quickly and, sometimes, big ones. Adaptive muddling (chapter 22) is well suited to such demands.

The adaptive-muddling principle builds on the small-experiment approach, which is a framework for proliferating solutions while maintaining local relevance and experimental validity, all the while promoting rapid dissemination of findings. It allows ordinary people to learn what options work in their community. Small experiments are going on all the time (e.g., pilot programs, field tests, demonstration sites, trial runs) but their effectiveness can be enhanced by following a few simple guidelines.[16] The aim is to identify reasonable solutions, not search for the ideal answer. Information, as in any experiment, is gathered and evaluated but not necessarily in a formal way because the intent is to enhance exploration rather than document findings for publication. Small-scaled explorations are well suited to localization since they lead to greater public engagement. And a key element is wide dissemination of results so that other communities can quickly benefit; successes in one locality become plausible options to explore elsewhere. It is also noteworthy that nothing in these guidelines restricts small experiments to small steps or to a slow discovery process. In fact, by making the experimental process less intimidating, more discovery can occur quickly as more people become engaged.

10. *Prefamiliarization principle* Localizers should consciously build and share mental models of positive localization. In conversations about behavior change, it is often mentioned how readily people anchor to the status quo and how immune they are to abstract scientific evidence.[17] One might think that this would pose a serious problem for localization efforts. After all, we may be called on to make far-reaching changes, away from the status quo, toward an unfamiliar life pattern, all based on abstract arguments.

However, the issue here may not be a status quo bias but a familiarity bias. A familiarity bias is based on our mental model of a situation and thus mirrors the strengths and weaknesses of our current understanding. This provides hope since mental models can be formed and altered in a large variety of ways.

Some adjustment to mental models can be gained through indirect personal experience. Consider the powerful effect of reading fiction, poetry, and memoirs, and of observing alternative living patterns (e.g., living farms, museums). Direct experience is equally effective at building and changing cognitive understanding (see the walkabout principle below). Consider the role of transition town workshops, farmers' markets, and local business initiatives in providing direct exposure to examples of localization that already exist in one's community. An even deeper experience can be had through, for instance, working on a farm or living temporarily in a materially simple but socially rich community.

11. *Tentative localization principle* Psychological preparation will be enhanced if people have the opportunity to try out portions of new life patterns without having to adopt the whole thing at once. People thinking about transitioning to a simpler life will benefit from first exploring subparts. Learning at one's own pace is more agreeable and avoids grave errors. This desire to try things out ahead of time is an essential component of effective adaptation. One should never assume that everything will be as first envisioned; experimentation, once again, is essential.

12. *Walkabout principle* While localized economies and the communities that constitute them will focus much of their productive energies internally, localizers should at the same time encourage members to explore, whether it's going to the next community over or around the world. Such exploration should be institutionalized.

Under a localization regime, insularity will remain a recurrent threat. Successful communities will have heightened internal connectivity, in contrast to the dissipated relations of the current globalized regime. The built world—residential, commercial, governmental—will, due to declining net energy and material availability, necessarily be designed for human-scale transportation (e.g., walking, bicycling, carting).[18] This will facilitate direct person-to-person interaction. From high-density housing to public squares to community gardens, farmers' markets, and community kitchens, members of the local community will necessarily interact regularly and frequently. The upside will be greater community cohesion. The downside will be a sense of isolation from the restricted mobility, not to mention internal, possibly growing, conflict.

To counteract the potential for insularity, then, localizers must institutionalize exploration beyond the community. Notice that this is still localizing. The direction of inquiry is still inward (toward local and regional integration), not outward (toward global integration), even though the physical movement and intellectual engagement are outward.

The walkabout principle is named after Australia's custom of sending young people out into the world after schooling and before settling down to work and family. Under localization, it could apply to all ages since its benefit to the individual is broadening horizons and its benefit to the community is acquiring new ideas, even new blood (some sojourners will no doubt stay away; others may return with new friends).

An Ongoing Conversation

Downshift is one of many useful metaphors in the transition out of a fossil fuel–driven, consumerist society and into a life of less material wealth, more time, better food, and enhanced well-being. Downshift, unlike collapse, implies choice: we can choose our path, we can decide to make it peaceful, democratic, just, and resilient, or not. Downshift, unlike collapse, also implies that things are under control. We decide to change because we are members of communities seeking to thrive. We explore alternatives, some marginal, some radical, without dread of the consequences of failure, with a sense of deep community support.

Finally, downshift implies an appropriate pace of life. We voluntarily decide how far to shift down and then work to figure out how we're going to live well at the slower pace. In fact, unlike the consumerist view that time is money and faster is better, the downshift view is that some

The Potato

If localization should ever acquire an icon, it might just be the potato. Some foods are inherently global, or at least they are bred, grown, packaged, and shipped in ways that make them global commodities. Wheat, rice, and corn are prime examples (not to mention sugar and oils). In contrast, other foods have their place and seem to stay put. Take the potato.

Potatoes are heavy and bruise easily, so they don't ship well, not over long distances anyway. What's more, they can be grown just about anywhere by anyone. Cut a potato in pieces, each with an eye, and stick them in the ground. In a few months they bear large starchy tubers. Each tuber is rich in protein, starch, vitamins, and nutrients like zinc and iron. Potatoes require less energy and water to grow than wheat. Pound for pound they contain less protein than wheat but an acre of potatoes yields more protein than an acre of wheat.

The United Nations and other international bodies are beginning to take notice, especially as a means of averting hunger in poor countries. But as cheap energy disappears, all of us soon may be throwing a few pieces in the ground, at least as backup. "When you plant a potato it gives you food security," says Dr. Pamela K. Anderson, director of the International Potato Center, a scientific research center in Lima, Peru. Unlike food aid, when local farmers grow potatoes it "strengthens the local economy, instead of just sending food." So, good for growers and eaters, good for the community's economy: localization's icon might just be the potato.
Source: Elisabeth Rosenthal, "Rearranging Pantries, Aid Groups Favor Potato," *New York Times*, October 26, 2008.

intermediate speed is healthier—healthier for creating strong communities, for living well and within our means, for enhancing personal well-being and societal integrity.

Clearly, we need new ideas, innovative designs, creative principles, and better language. We need better ways of expressing the imperative of living well with less. As editors, we hope this book prompts a conversation about downshifting, not about collapsing. We hope it spurs more such efforts as localization proceeds.

Notes

1. Thomas Homer-Dixon, The Upside of Down: Catastrophe, Creativity and the Renewal of Civilization (Toronto: Knopf, 2006).

2. Curiously, our conversations with people in industry, including the oil industry, and the military, suggest that these sectors at a minimum do find the scenario laid out here plausible. It is the mainstream policy communities—politicians, the mainstream media, environmentalists—who find this scenario difficult to accept.

3. Herman E. Daly, "Sustainable Growth: An Impossibility Theorem," in Herman E. Daly and Kenneth N. Townsend, eds., Valuing the Earth, 267–273 (Cambridge, MA: MIT Press, 1996) (quote on 267).

4. Jerry Mander, In the Absence of the Sacred: The Failure of Technology and the Survival of the Indian Nations (San Francisco: Sierra Club Books, 1999), 190.

5. Quoted in Kirkpatrick Sale, "The Decentralist Tradition," in Kirkpatrick Sale, Human Scale, 443–454 (New York: Perigee Books, 1980) (quote on 446).

6. David C. Korten, The Great Turning: From Empire to Earth Community (San Francisco: Berrett-Koehler, 2006) (quote on 293); Thomas Princen, The Logic of Sufficiency (Cambridge, MA: MIT Press, 2005) (quote on 278).

7. Mary Midgley, "Toward a New Understanding of Human Nature: The Limits of Individualism," essay prepared for the Seventh International Smithsonian Symposium, "How Humans Adapt: A Biocultural Odyssey," November 8–12, 1981, Smithsonian Institution, Washington, DC, 1–24 (quote on 18).

8. Stephen Kaplan, "Human Nature and Environmentally Responsible Behavior," Journal of Social Issues 56, no. 3 (2000): 491–508 (quote on 498).

9. Wendell Berry, "The Idea of a Local Economy," Orion, Winter 2001, 28–37.

10. John Cavanagh and Jerry Mander, Alternatives to Economic Globalization: A Better World Is Possible, 2nd ed. (San Francisco: Berrett-Koehler, 2004).

11. Berry, "The Idea of a Local Economy," 37.

12. On the public trust doctrine in the context of water, see Zygmunt J. B. Plater, Environmental Law and Policy: Nature, Law, and Society, 3rd ed. (New York: Aspen Publishers, 2004).

13. We thank Alan Bush for bringing this example to our attention.

14. Quoted in Kurt Cobb, "We Must Make a Lot of Mistakes Quickly," Resource Insight, March 14, 2009. Retrieved from http://resourceinsights.blogspot.com/2009/03/we-must-make-lot-of-mistakes-quickly.html.

15. James C. Scott, Seeing Like a State: How Certain Schemes to Improve the Human Condition Have Failed (New Haven, CT: Yale University Press, 1998) (quote on 345). Also note his other guidelines, all of which are useful for localization: favor reversibility, plan on surprises, and plan on human inventiveness (p. 345).

16. Katherine N. Irvine and Stephen Kaplan, "Coping with Change: The Small Experiment as a Strategic Approach to Environmental Sustainability," Environmental Management 28 (2001): 713–725.

17. Alan AtKisson, Believing Cassandra: An Optimist Looks at a Pessimist's World (White River Junction, VT: Chelsea Green Publishing, 1999).

18. Ivan Illich, Energy and Equity (New York: Harper & Row, 1974).

Contributors

Gar Alperovitz, Department of Government and Politics, University of Maryland, is author of *America Beyond Capitalism: Reclaiming Our Wealth, Our Liberty, and Our Democracy* (2005).

Sharon Astyk, author of *Depletion and Abundance: Life on the New Home Front* (2008), is a writer and farmer in upstate New York.

Wendell Berry, a writer and farmer in Kentucky, is author of *Home Economics* (1987).

Adam Dadeby, is a business and energy analyst in the United Kingdom and cofounder of Passivhaus Homes.

Raymond De Young, School of Natural Resources and Environment, University of Michigan, is author of "Some psychological aspects of a reduced consumption lifestyle," *Environment and Behavior* (1996).

John S. Dryzek, School of Politics and International Relations, Australian National University is author of *Deliberative Democracy and Beyond* (2000).

David J. Hess, Sociology Department, Vanderbilt University in Tennessee, is author of *Localist Movements in a Global Economy* (2009).

Rob Hopkins, founder of the Transition Network in the United Kingdom, is author of *The Transition Handbook: From Oil Dependency to Local Resilience* (2008).

M. King Hubbert, was a geophysicist at Shell Oil Company and professor at Stanford and University of California Berkeley, as well as author of numerous papers on peak oil.

Ivan Illich, author of *Energy and Equity* (1974), was a visiting professor at Penn State and the University of Bremen.

Warren Johnson, Department of Geography, San Diego State University, is author of *The Future is Not What it Used to Be: Returning to Traditional Values in an Age of Scarcity* (1985).

Rachel Kaplan, School of Natural Resources and Environment, University of Michigan, is coauthor of *With People in Mind: Design and Management of Everyday Nature* (1998).

Stephen Kaplan, Department of Psychology, University of Michigan, is coauthor of *Cognition and Environment: Functioning in an Uncertain World* (1983).

Karen Litfin, Department of Political Science, University of Washington, is author of *Ozone Discourses: The Politics of Science in International Environmental Negotiations* (1994).

Thomas A. Lyson, Department of Development Sociology, Cornell University in New York, was author of *Civic Agriculture: Reconnecting Farm, Food, and Community* (2004).

Dennis Meadows, formerly the director of the Institute for Policy and Social Science Research at the University of New Hampshire, is coauthor of *The Limits to Growth* (1972).

Donella H. Meadows, was the founder of the Sustainability Institute in Vermont, wrote a weekly online column called "The Global Citizen," and was lead author of *The Limits to Growth* (1972).

Lester W. Milbrath, directed the Research Program in Environment and Society at the University of Buffalo-SUNY and was author of *Envisioning a Sustainable Society: Learning Our Way Out* (1989).

Thomas Princen, School of Natural Resources and Environment, University of Michigan, is author of *The Logic of Sufficiency* (2005).

Jørgen Randers, the Norwegian School of Management, is coauthor of *The Limits to Growth* (1972).

Josiah Royce, University of California, Berkeley, and Harvard University, was author of *Race Questions: Provincialism and Other American Problems* (1908).

Kirkpatrick Sale, a freelance writer in Vermont, is author of *Human Scale* (1980).

Ernst F. Schumacher, founder of the Intermediate Technology Development Group, wrote *Small Is Beautiful: Economics as if People Mattered* (1973).

Michael Shuman, director of research and public policy at the Business Alliance for Local Living Economies in Takoma Park, Maryland, is author of *The Small Mart Revolution: How Local Businesses Are Beating the Global Competition* (2006).

Joseph A. Tainter, Department of Environment and Society at Utah State University, is author of *The Collapse of Complex Societies* (1988).

Robert L. Thayer, founded the Landscape Architecture Program at the University of California, Davis, and is author of *LifePlace: Bioregional Thought and Practice* (2003).

Index